COLORADO'S SPANISH PEAKS REGION

AN EXPLORATION GUIDE TO HISTORY,
NATURAL HISTORY, TRAILS, AND DRIVES

RICHARD C. KEATING

COLORADO'S
SPANISH PEAKS REGION

AN EXPLORATION GUIDE *to* HISTORY, NATURAL HISTORY, TRAILS, *and* DRIVES

ISBN 978-1-930723-85-6
Library of Congress Control Number 2010932694

Scientific Editor and Head: Victoria C. Hollowell
Managing Editor: Beth Parada
Associate Editor: Allison M. Brock
Associate Editor: Tammy M. Charron
Press Coordinator: Cirri R. Moran

Design and typesetting by Adrianna Sutton.
www.adriannasutton.com

P.O. Box 299
St. Louis, Missouri 63166-0299, U.S.A.
www.mbgpress.info

Dedicated to my wife, Nancy Jodean Keating

CONTENTS

PREFACE

DURING OUR first summer in this area, my wife, Jody, and I looked in vain for a compact introductory guide. We bought a trail guide to the Sangre de Cristos that claimed to be comprehensive for this area. Its virtual non-coverage of the Spanish Peaks left us with a blank spot in the literature that begged to be filled.

Once, I expressed my admiration to a colleague who had written fourteen texts introducing various branches of mathematics. Writing a book on a new topic, he explained, was his response to complete ignorance. I found that inspiring. Certainly, in this case, I have not written a volume that a native could have, but I have enjoyed having the chance to break this ground.

The original purpose of this project was to write a portable source that could find a home in a glove compartment or daypack. Early on, as facts and ideas for exploration multiplied, the project began to look increasingly encyclopedic. The guide remains incomplete, but this seems to be a good time to allow it to hatch.

Unlike conditions in more urban areas, this region presents an ideal settlement pattern. Great tracts are held in common in perpetuity, a circumstance that invites everyone to partake of its soft or stark moods, its sometimes unforgiving climate, its complexity, surprises, and grace. Heed the cry of a landscape that cries out to be explored.

Disclaimer: The whole purpose of this volume is to be accurate, informative, and useful. However, there is no way that data and descriptions can be guaranteed in an ever-changing world. There are also inherent risks in all forms of travel. The author and publisher can assume no responsibility for accidents or misjudgments consequential to use of this guide.

Occasionally, businesses are mentioned as sources or as landmarks. No special endorsement is intended over those not mentioned.

ACKNOWLEDGMENTS

THE LIST of people who have helped advance this project is long. It runs the gamut from those who have provided extensive information, or a valuable lead, to those who provided an interesting gem in a casual conversation. I am grateful to former colleagues Fred and Marion Lampe for their friendship, for introducing Jody and me to the area, and for the valuable lore they have passed our way. We are also thankful that they supported the reprinting of the valuable *History of the Spanish Peaks Ranger District* written by Lewis Cummings in 1947.

Former colleagues and friends Ellie and George Mellott have passed along their keen interest and have introduced us to numerous folks who know and love this region. Tours once sponsored by the late Walsenburg center of Trinidad Junior College, including those by local historians David Steffan and his former teacher Betty Story, were well prepared and informative.

Through enjoyable conversations, or through tours of museums and historical sites, we have learned of the region's history and geography in ways that go far beyond the surface. I am especially grateful to Nancy Christofferson, Tom Doerk, Katharine Emsden, Jack Estes, Carolyn Newman (as Mother Jones), Betty and Verne Story, Gene Vories, and Bill Williams. Loretta Martin, director of the Louden-Henritze Archeology Museum at Trinidad State College, was very generous with her time. I've gained much from the signature columns of Jim Conley and Bob Kennemer. We are privileged indeed to have libraries in La Veta, Walsenburg, and Trinidad staffed by librarians who really care about the region and its culture, and want everyone to love it as they do.

On the natural history side, I am grateful to Neil Snow, formerly director of the Herbarium at the University of Northern Colorado, for his thorough and expert review of the plant list for the region, and to George Yatskievych and Jim Zarucchi for advice on various issues that arose while I edited checklists. Mick Richardson commented on the bird list that needed much help.

Phil Keating offered a valuable critique of the ecology section from the standpoint of a physical geographer. Katharine Emsden and Ellen Kunkelmann advanced the project with their criticism of earlier drafts. The book took its final shape through the thorough attention given by Allison Brock and Beth Parada, editors at the Missouri Botanical Garden Press. Gordon Keating, geologist, coauthored the geology chapter. Try as I do to keep abreast of modernity, my original training occurred before continental drift was considered believable.

In the end, I made judgment calls and am therefore responsible for errors and omissions. I will be glad to have them brought to my attention. Characteristic of a special place is the willingness of new friends and near strangers to discuss their beloved countryside when they believe you are really interested. Their unqualified love of this landscape is obvious and infectious, and I am grateful to all of them.

CHAPTER 1

REGIONAL SETTING AND INTRODUCTION

WHEN APPROACHED from any direction, the Spanish Peaks stand out as dramatic focal points. The surrounding peaks, passes, valleys, and plains encompass a region of unique history and natural history. From the Spanish Peaks one can see three rivers, the Huerfano, the Cuchara, and the Purgatoire, each of which begins as a rivulet on the flanks of the Sangre de Cristo Range. The ridgeline and these three basins form the natural geographic limits on the north, west, and south. The edge of the Great Plains, roughly following I-25, forms a natural boundary on the east.

As usual, political boundaries tend to be more arbitrary, and this otherwise natural scene is bisected by pieces of two counties. Las Animas County's northeastern border follows the irregular ridgeline of the Spanish Peaks to its crossing at I-25. To the west, the county line is traced across Cuchara Pass to Trinchera Peak, a corner that separates these two counties and Costilla County of the San Luis Valley to the west.

From Trinchera Peak, the western Las Animas County line follows an arbitrary straight line south to the New Mexico boundary. On the Great Plains, the county boundaries are the usual straight lines that surveyors are wont to draw across treeless and lightly populated regions.

The Spanish Peaks rise dramatically 6000 to 7000 feet above their bases and are all the more arresting for their relative isolation. In fact, they have nothing to do with the forces that elevated any neighboring ranges. For their height class, they are the easternmost members of the Rocky Mountains.

HUERFANO COUNTY

All of Huerfano County west of I-25 is within this guide's region. In the 19th century, travelers arriving from the plains were greeted by a small volcanic butte near the crossing of the Huerfano River. They named it *Huerfano*, "the Orphan," and so the county has worn that name ever since. The Spanish Peaks, the adjacent Huerfano River Valley, and the Wet Mountains to the north formed distinctive features signaling that Spanish territory was heaving into view. This trail along the mountains became the Mountain Route of the Santa Fe Trail. For a brief time, in the days of thin population, Huerfano County was the largest in Colorado.

LAS ANIMAS COUNTY

Only the northwest corner of this large county includes the Spanish Peaks neighborhood. The Purgatoire Valley from the headwaters to the city of Trinidad forms a natural boundary.

RIVERS

The Huerfano River rises in the northeast-facing basin of the four Blanca peaks, each of which exceed 14,000 feet and which constitute the southernmost rampart of the northern Sangre de Cristo Range. This range defines unambiguously the western margin of Huerfano County. Gathering its numerous tributaries in a broad agricultural valley, it flows northeast past Gardner, then east past the southern portion of the Wet Mountain Valley. It travels through Huerfano Park at the foot of the Wet Mountains, and then after crossing I-25, it turns northeast, entering the Great Plains at its junction with the Arkansas River near Pueblo. Closer to its headwaters, English settlers corrupted the Spanish name from Huerfano to "Wharf" Creek and then to "Wolf" Creek. A major tributary, Pass Creek, has its headwaters at a low divide just east of La Veta Pass. Road 372 follows Pass Creek and connects the Huerfano Valley to Route 160.

North of Route 160, a low east-to-west ridge separates the Cuchara River Basin to the south. Its valley is narrower, leaving less room for agriculture. The river originates in the snow field basins of the flank between Trinchera Peak and Teddy's Peak of the southern Sangre de Cristos, or Culebra Range. From Bear and Blue lakes, it travels north down the narrow picturesque Cuchara Valley, roughly parallel with Route 12. It turns east at La Veta and, after reaching the plains, becomes a tributary of the Huerfano.

Continuing south on Route 12 over Cuchara Pass Divide, one descends into the Purgatoire River headwaters region. The river's main stem originates from a collection of streams from adjacent parallel valleys of the Culebra Range. On its way toward Trinidad, the Purgatoire is confined by a narrow valley, with the Spanish Peaks and Cuchara Pass to the north and a range of hills leading to New Mexico to the south. Flowing eastward, the river passes through a mix of agricultural and historical coal-mining neighborhoods before entering into Trinidad reservoir. Beyond Trinidad, it flows northeast through gentle canyon country on its way to the Arkansas River.

One feature of western rivers strikes easterners as odd: as these rivers go downhill, picking up tributaries, water volume decreases. There seem to be two reasons for this phenomenon. In arid lands the ground is seldom fully saturated. However, flow reduction is due mostly to removal of water at various points during its travel—an exercise of irrigation rights. Travelers from well-watered places are amused by the hopeful use of "river" for bodies of water they would call creeks or streams. But in parched country, it seems normal to endow this crucial resource with a certain nobility.

Another interesting feature of these river valleys is their geographic isolation. Over the past 150 years, valley settlers and their descendants have developed distinct cultures and histories. To this day, geographic barriers tend to preserve this variety, which is comforting in an era that tends toward uniform popular culture.

MOUNTAINS

There are three large elements and many smaller ones. Beginning in the north, the Wet Mountains form a large mass looming over the Huerfano Valley. Greenhorn Peak, at the range's south end, is its only distinct elevational feature. The north-northwest trending range is connected to elements of the Front Range at its north end. Traveling toward Gardner on Route 69, one sees the range's south end rising abruptly from the valley. The uplift that raised the mass was controlled by faults on the east, south, and west sides.

West of the Wet Mountains is the Wet Mountain Valley, which, in turn, ends abruptly at the Sangre de Cristo Range. The "Sangres" form the longest range in Colorado and one of the longest ranges on the continent. In the north, the Sangres begin at the end of the Sawatch Range and stretch from Salida in central Colorado to Glorieta Pass near Santa Fe, New Mexico, a distance of more than 220 miles. One ridge wide, the range forms an almost continuous wall that divides this region from the San Luis Valley to the west.

Along Route 160, between La Veta Pass and the town of Blanca, there is a misalignment between the northern and southern Sangres of about 25 miles.

Nevertheless, geologists consider the two sections to have been elevated by the same forces and to be one contiguous range.

To really understand a landscape, one must go beyond knowing lists of names or catalogs of events and dates. The meaning of landscape elements and the ways in which people adapt to them are only slowly understood. The good news is that sharpened awareness is available to any amateur naturalist. Small revelations and notes accumulate through time. Consider this guide a starter kit for this quest, a path that has no end.

There is wonder here. When the trail becomes narrow and indistinct, you face a landscape that has nothing to do with man. It is awe inspiring to discover that civilization ends here and the world remains indifferent to our presence. We enter the home space of fellow species, remarkable survivors of eons on the surface of an earth where change is the only constant.

CHAPTER 2

FEDERAL LANDS

SAN ISABEL NATIONAL FOREST

The San Isabel National Forest, one of 11 national forests in Colorado, comprises four noncontiguous tracts. The smallest and southernmost unit includes the Spanish Peaks and adjacent Sangre de Cristo Range. Its regional office is in La Veta. The San Isabel Forest Reserve was established in 1902 and became the San Isabel National Forest in 1907, soon after the founding of the U.S. Forest Service within the Department of Agriculture. Between 1907 and 1945, acreage increased as several other forests were integrated into it. The San Isabel now includes about 1 million acres and varies in elevation from 5860 feet on the plains to 14,433 feet at the summit of Mount Elbert, Colorado's highest fourteener. Seven wilderness areas are designated within the San Isabel.

According to the Wilderness Act of 1964, a wilderness is "an area where the earth and its community of life are untrammeled by man and where man himself is a visitor who does not remain." Visitors are requested to practice "leave no trace" techniques. At a minimum, this means that you should pack out your refuse and leave as little impact from your visit as possible. Use packable camp stoves rather than burning wood fires. Familiarize yourself with wilderness regulations by contacting the regional agency office in advance of your trip (see Resources). Three wilderness areas are within our guide range.

SPANISH PEAKS WILDERNESS

Of the nation's 662 dedicated wilderness areas, 23 were signed into law in 2002 by President Clinton. Among those was Colorado's newest wilderness area, centered on the summits of the East and West Spanish Peaks. It includes about

19,200 acres. The area was first placed on the National Registry of Natural Landmarks in 1977 and was recommended by the U.S. Forest Service as suitable for wilderness designation in 1979. While there was broad support for the idea that this superb scenic and natural resource deserved protection as wilderness, a lengthy period of public involvement and hearings followed, especially regarding the fate of inholdings.

U.S. Senator Wayne Allard (R-CO) sponsored the Senate bill, and Representative Scott McInnis (R-Grand Junction) shepherded the House version through Congress. On October 23, 2000, Congress passed the bill approving the dedication of the Spanish Peaks section of the Sangre de Cristo National Forest as a protected wilderness area. Soon after, on November 7, 2000, President Clinton signed the Spanish Peaks Wilderness Act into law. The long-awaited dedication ceremony occurred on Saturday, September 9, 2001, at Cordova Pass.

Senator Allard said the dedication marks "...the completion of a dream for many of us who have worked for decades to see the magnificent twin towers of the Spanish Peaks protected for future generations." Representative McInnis noted that "These twin peaks are majestic symbols of Colorado's natural character that eminently deserve the permanent wilderness safeguards they have received...They are spectacular treasures that have been guaranteed protection for all time."

Translated from the Spanish, Las Cumbres Españolas gives us the present name, the Spanish Peaks. Among names used in earlier times were Dos Hermanos ("two brothers"), Mexican Mountains, and Wahatoya, which seems to have two interpretations. Often, it is said to mean "breasts of the world" in the Ute language. But according to the late John Brubaker, amateur student of Plains Indian linguistics, the name actually means "twin peaks" from a Comanche word that translates as *waha*, meaning "two" or "twin," and *toya*, meaning "mountain." Brubaker's foray into Indian linguistics is housed in the La Veta Public Library. The Indian name, also spelled Huajatolla or Guajatoyah, is still fixed to various landmarks in this region.

A regional overview of the Spanish Peaks is efficiently made on a loop drive. Beginning at Walsenburg, take Highway 160 to Highway 12, go south over Cuchara Pass to Trinidad, and return to Walsenburg via I-25 north. The loop can also be bisected for a closer approach to the peaks. At Cuchara Pass on Highway 12, take the Apishapa gravel road east, which ascends to Cordova Pass and then descends to Aguilar and I-25. It passes to the south of the summits, which can be seen from several vantage points (see Chapter 7, Drives).

Because of distorted perspective in many places, it is hard to see any difference in summit elevations. Actually, West Peak (at 13,626 feet)* is about 1000 feet higher than East Peak (at 12,708 feet). However, East Peak is a more challenging climb because of the difficulty of approach. Although a fairly high trailhead leaves Cordova Pass for West Peak, there is no high-elevation automobile access to East Peak. It is well embedded within wilderness (see Chapter 8, Trails).

You should note that the various national forest roads occasionally pass through private property. Watch for postings and do not enter these properties without permission. (See Appendix 2, as well as Chapters 6 to 8 on Geology, Drives, and Trails, for more detailed information.)

GREENHORN MOUNTAINS WILDERNESS AREA

To the north, at the summit of the Wet Mountains, Greenhorn Peak looms over Huerfano County. This wilderness area comprises 22,040 acres and ranges in elevation from 7600 to 12,300 feet. It was dedicated in 1993 and remains one of the more remote and seldom-visited wilderness areas in Colorado. There are no jagged peaks or lakes, and only 11 miles of trails are found in the northern portion. However circuitous, the trip to Greenhorn Peak is worth the effort (see Chapter 7, Drives). A gravel road leads to a parking area at the wilderness boundary, and trails lead to North Peak (12,221 feet) and South Peak (12,346 feet). In addition to the spectacular views of Huerfano County and beyond, there are superb wildflower displays, wonderful crystalline rock formations, and ancient trees.

The area's southern zone has no trails, and it gets drier as the elevation drops. Exploration would require bushwhacking through steep and irregular terrain. Winter winds sweep many slopes clear of snow and thus furnish an important winter range for bighorns, mule deer, and elk.

SANGRE DE CRISTO WILDERNESS AREA

This area is administered by the San Isabel National Forest (see Resources)

*On the West Peak summit in 2006, our party, working with three GPS units on a clear day, pulled in nine strong satellite signals. We came up with an elevation of 13,563 feet. Considering the rate of erosion of the summit mass, it is not surprising that the figure may be somewhat lower than that officially measured years ago.

and became a dedicated wilderness in 1993. At 226,455 acres it is the third-largest wilderness in Colorado. Extending about 70 miles in length, it covers the whole northern Sangre de Cristo Range from near Salida at the northern terminus and south to include Mount Blanca. The ridgeline ranges in elevation between about 8000 and 14,294 feet and includes several of Colorado's fourteeners.

The origin of the name Sangre de Cristo is lost even to legend. Was it given for the color of the sunset that immerses the peaks? Or was it due to the cry of a dying martyred Spanish priest?

GREAT SAND DUNES NATIONAL PARK AND ENVIRONS

This area in the eastern San Luis Valley, hard against the western flank of the Sangre de Cristo Range, is a significant complex of several contiguous properties. They are managed by a partnership of private foundations and state and federal agencies. Land and water rights are protected intact (Fig. 1).

GREAT SAND DUNES NATIONAL PARK AND PRESERVE

Protection was originally established by legislation signed into law in 1932 by

Fig. 1: Medano Creek at dusk, looking southwest toward the Great Sand Dunes.

President Hoover with the goal of saving the unique dunes from various forms of exploitation occurring in the early years of the century. The impetus began as a movement of San Luis Valley citizens, spearheaded by the leadership of Philanthropic Educational Organization (PEO) chapters, a women's organization.

In the intervening years, various additions have greatly expanded protection of this significant ecosystem. On November 22, 2000, President Clinton signed into law a statute enabling the monument to become a national park when the Secretary of the Interior completed negotiations for certain planned land acquisitions. In 2005, Secretary of the Interior Gale Norton completed these tasks and dedicated the national park.

GREAT SAND DUNES WILDERNESS

This wilderness, the 33,450-acre center of this stunning national park, was dedicated in 1976. It constitutes what biologists consider to be a fine example of a saltbrush-greasewood ecosystem. The formidable 700-foot dunes are the largest in North America, and their position at the very foot of the steeply ascending Sangre de Cristos creates beautiful visual contrasts, especially at low

Fig. 2: A wetland adjacent to Great Sand Dunes National Park. It is bounded by dunes in the background, with the northern Sangre de Cristo Range on the skyline. The low point on the left horizon is Medano Pass.

solar angles (Fig. 2).

In the wilderness, camping is allowed but not campfires. To explore the area, obtain a free back country permit from the park visitor center on the main road inside the park entrance. To reach the Great Sand Dunes, take Route 150, which branches north from US 160 just west of Fort Garland in the eastern San Luis Valley (Fig. 3).

BACA RANCH

A few years ago the Nature Conservancy spent more than $31,000,000 to buy the 97,000-acre Baca Ranch, which borders the Great Sand Dunes National Park on the northwest. The 151-square-mile property ranges in elevation from 7500 to 14,165 feet at the top of Kit Carson Peak. Complex negotiations between the Department of Interior, the Nature Conservancy, and the Colorado State Land Board led to the transfer of the property to the Great Sand Dunes Park before it became a national park.

MEDANO-ZAPATA RANCH

Adjacent to the south of the Great Sand Dunes National Park is the final addi-

Fig. 3: The Great Sand Dunes. These dunes are the tallest in North America due to unusual circumstances. Medano Creek carries a continual load of sand windward. In addition, annual changes in wind direction build the dunes from the southwest and the northeast.

tion to this complex. In 1999, the Nature Conservancy obtained the 103,000-acre Medano-Zapata Ranch (once two separate ranches) from private owners. This property will continue to be managed by the Nature Conservancy.

From Route 150, the ranch appears to be uninteresting scrub desert, but on close inspection it reveals much interesting ecology. Not only does the tract have great elevational diversity, but it also has a high water table that includes ponds around the base of the dunes. As a result, more than 3500 species of plants exist on the ranch properties alone. This is huge. At least 70 rare and imperiled species in all categories occur here—and some only here. Old fences are being removed, and about 2000 bison will continue to graze on 15,000 acres of free range (Fig. 4). This brings the total contiguous protected land in the best biological hot spot in the San Luis Valley to nearly 300,000 acres. Contact the Nature Conservancy of Colorado for information regarding tours of the Medano-Zapata Ranch.

SOUTHERN SAN JUAN WILDERNESS AREA

Strictly outside this guide's range to the west, this wilderness forms the west-

Fig. 4: A bison herd grazing on Medano-Zapata Ranch, south of and adjacent to Great Sand Dunes National Park.

ern rampart of the San Luis Valley. It was dedicated in 1980, was augmented in 1993, and has a colorful history. At 185,790 acres, the Southern San Juan Wilderness is one of Colorado's largest forests and ranges in elevation from 8000 to 13,000+ feet. It includes land administered by the Rio Grande and San Juan national forests. For more than 50 miles, the wilderness straddles the Continental Divide and has become a favorite backpacking destination.

It is also identified as the Tierra Amarilla Land Grant tract, and it remains wild and nearly impenetrable. Of special note is that the area is the site of the "last" grizzly bear encounter in 1979. Wildlife biologists agree it would be the best place remaining in Colorado to find grizzlies if they still survive in southern Colorado.

DINOSAUR TRACK SITE

Also outside the guide's area to the east, an exploration of the canyon and track site is worth the extra miles. The remote and wild canyon of the Purgatoire River in Las Animas County displays a spectacular collection of 1300 dinosaur tracks in about 100 different trackways. The giant footprints were originally laid down 150 million years ago on shoreline muds of the Jurassic period by allosaurs and brontosaurs, and they provide evidence of family groupings not seen elsewhere. On the long walk in from the north (Figs. 5 and 6), one can find ancient Indian rock art, probably less than 2000 years old, and the remains of the picturesque Delores Mission and Cemetery, which were built between 1871 and 1889 by Mexican pioneers.

The track site is within the Picketwire Canyonlands tract, which was transferred to the Comanche National Grassland, administered by the U.S. Forest Service, in 1991. Your first stop should be the Comanche National Grassland headquarters in La Junta. Personnel there will provide details on current hiking conditions and tour availability. Access to this primitive site (there is no camping) is from La Junta, Colorado, using Highway 109 south (see Chapter 7, Drives).

GETTING INFORMATION AND HELP

U.S. FOREST SERVICE

Forest boundaries are well marked on trails and roads, and rules regarding the use of trails are clearly posted. Most popular trailheads are called *fee areas*, where visitors deposit an envelope with cash into a slotted pipe. Finding an informed

Fig. 5: Purgatoire Canyon near the north end of the trail system. Note the concentration of vegetation near the river.

Fig. 6: Typically gnarled old plains cottonwood tree in the Purgatoire Canyon south of La Junta.

employee in the field is another matter. I've never encountered a patrol ranger in 40 years. Fee collection personnel and campground hosts are private contractors who usually have little information outside their assigned tasks.

At the district office in La Veta, you can buy a forest map and find a brochure rack. For detailed information, your best bet is to visit the headquarters in Pueblo (see Agencies and Organizations, Resources), where you can find informed specialists who are glad to take the time to help you. Knowledgeable forest staffers are professionals whose responsibilities are large and whose resources are stretched to the limit. Given this stress, their public service duties are reasonably well met.

BUREAU OF LAND MANAGEMENT (BLM) LANDS

In this area, BLM lands are scattered around in discontinuous parcels, sometimes at the edges of the National Forest and sometimes not. There is little signage indicating where these boundaries are. For information, one should check with the nearest BLM office. One is in the Royal Gorge area to the north, which covers the Spanish Peaks region. The La Jara field office is in the San Luis Valley to the west (see Resources).

STATE WILDLIFE AREAS (SWAs)

The Colorado Division of Wildlife manages 230 properties with a total of 369,518 acres and leases an additional 481,333 acres. Its purpose is protection of habitat for wildlife, plus activities such as hunting, fishing, hiking, photography, and wildlife watching as appropriate. Hunting and fishing must be done in accordance with state license regulations that vary from area to area.

As of December 2010, the $10.25 Colorado Wildlife Habitat Stamp must be obtained to visit all state wildlife areas. It cannot be obtained on site, but a $68 fine can be assessed for failure to possess this permit. All visitors must buy one, although the permit is free for Colorado residents younger than 18 or older than 65 years of age. Wildlife Habitat Stamps are available where hunting and fishing licenses are sold. In the present tax climate, many management needs are being met more frequently by assessing user fees.

Huerfano SWA

This Huerfano County site is 13 miles west of Gardner on County Road 580, southwest of Sharpsdale. It comprises 544 acres at an average elevation of 7600 feet and lies astride the Huerfano River near its headwaters. This section is prime habitat for four species of trout. Hiking and camping as well as hunting and fishing can be enjoyed.

Bosque del Oso SWA

This large, 33,000-acre tract is in Las Animas County south of Route 12 in the Purgatoire Valley. The land elevation ranges from 7200 to 8800 feet. One entrance is at Weston on the south side of the highway, and the other is just to the west on Route 12. Enjoy cold-water stream fishing, hunting, wildlife viewing, and photography.

Apishapa SWA

From Walsenburg, drive 18.5 miles east on Highway 10, turn right (south) on County Road 77, and drive 7 miles. At County Road 90, turn east and drive 11 miles to the property.

EXPLORATION AIDS

Stores that serve hunting and fishing needs in Walsenburg and Trinidad have selections of local relief maps, as does the visitor center store of Great Sand Dunes National Park. Complete sets of U.S. Geological Survey topographic maps are available at the federal building in Denver and can also be ordered by mail. Agency websites often have downloadable index maps. Google Earth offers additional maps. An internet search for topographic maps will turn up several firms that sell them online.

CHAPTER 3

BRIEF OUTLINE OF REGIONAL HISTORY

AMERICAN INDIANS

Archeological information for the upper Purgatoire Valley suggests that American Indian hunter-gatherers have occupied the region off and on for more than 3000 years. By 800 AD, there were longer-established camps in the valley where people hunted and engaged in growing corn as well as collecting pinyon nuts, prickly pear cactus fruits, and seeds from the nearby foothills. The Sopris culture occurred around 1025 AD and lasted for about two centuries. These people built larger stone and adobe pueblos and cultivated gardens in flat areas along the rivers. There is some evidence that severe drought caused the disappearance of bison and other large animals, and therefore people, from the southern plains. These conditions lasted until about the 14th century. After that time, we see the emergence of today's recognizable tribes, the Apaches, Utes, and Comanches, as described below.

The written history of European settlement and its impact on the American West in the last 400 years is reasonably well described in the available literature. Much less appreciated is that, even before European settlement, American Indian tribes were undergoing major changes in their ways of life. The acquisition of the Spanish horse after 1598 opened new possibilities that led to changes in territory size, hunting, agriculture, trading, and raiding. The Pueblo Revolt of 1680 accelerated the spread of horse culture in all directions. By the early 1700s, many tribes lived entirely differently than their ancestors due to greater mobility. Now they were able to defend larger territories and hunt with previously unimagined efficiency—the evolution of "hyper-Indians" as some have called them. But at the same time, between the 16th and 19th centuries, recurring cholera and smallpox epidemics from European contact decimated many tribes and bands. Population losses were as high as 95% in some areas.

UTES

From at least the first century AD, the mountains and adjacent valleys were Ute territory. The Utes lived at the highest elevations of any American Indians and occupied much of Colorado. Their Shoshonean language became separated from the Uto/Aztecan linguistic group probably about two millennia ago. The earliest Spanish Peaks name, *Wahatoya*, is said to be of Ute origin, but a similar word occurs in the related Comanche language.

At the time of European exploration and settlement, tribal members were dispersed among seven separate bands ranging over Utah and Colorado. Of these, the Mouache band occupied territory along the mountain front from about Denver to Las Vegas, New Mexico. The Tabeguache band (also called the Uncompahgre band) lived in an area that extended from present-day Pueblo, Colorado, to Taos, New Mexico, including the Spanish Peaks. The Capote band lived in the San Luis Valley.

Before obtaining horses, and in response to dispersed food resources, small family bands engaged in hunting and seed gathering from various camps. For projectile points, the Tabeguache band mined chert formations near Gardner. Some bands planted the well-known triad of New World domesticates: corn, beans, and squash. In the fall, the small bands moved out of the mountains to spend the winter in larger groups.

Although the Spanish settled New Mexico in 1598, their earliest reference to the Utes was about 1826. Within a generation, the arrival of the horse and trade goods changed Ute lives abruptly. Being able to move quickly made it possible to live together in larger numbers. They became raiders and fought with neighboring tribes over horses and livestock. The 18th century saw heightened conflict between the various tribes as well as with the Spanish. Treaties, hostilities, and friendly trade relations alternated. In their position at the edge of the Great Plains, they had territorial disputes with the Kiowa Apaches near present-day Walsenburg.

The last great Tabeguache chief was Ouray. Born in 1833 near Taos Pueblo, Ouray's education included learning to speak English, Spanish, Ute, and Apache. In 1860, following in the footsteps of his father, he became chief. He and his wife Chipeta were peacemakers and were fatalistic with respect to their chances against white dominance. Considered a polished negotiator, Ouray solved many conflicts through diplomacy during the remainder of his life. Surely the loss of their traditional ways was tragic for all western North American native peoples in the 19th century, but Ouray's wise and steady leadership is said to have eased this transition for his people.

COMANCHES

At the time of European settlement, the Comanche tribe occupied the plains to the east of the mountains. Beginning as eastern Shoshones, the Comanches were one of the first tribes in historic times to move from the intermountain region onto the Great Plains. In the early 1500s, they were hunting bison and using dog power to move their four-pole tepees east of the Laramie Mountains. This lasted a few generations and was made possible because of wetter climate conditions on the plains and consequently larger bison herds.

By the late 1600s, the Comanches were displaced back to the mountains and sagebrush by the Assiniboine and Blackfeet tribes that had recently acquired firearms. In turn, the great westward migration of Sioux people from the Great Lakes region displaced many tribes from the prime bison range of the northern Plains. The Comanches' dominant occupancy of the southern portion of the western Great Plains ended in 1779 when their last great chief, Greenhorn, was defeated by troops of the Spanish governor (see later discussion).

Ute and Comanche bands continued to occupy the area through 1881. Several thousand Utes continued to camp between Mount Maestas (then called Baldy or La Veta Mountain) and Silver Mountain. They also camped in the upper Cuchara Valley and along the Huerfano River. Periodic conflicts occurred between settlers and Utes and between Utes and Comanches before these bands were finally restricted to reservations.

APACHES

Relative latecomers to the region, the Apaches entered the Southwest and the upper Purgatoire Valley earlier than 1600 AD, by one estimate as early as 850 AD, and by other estimates around 1150 AD. Along with the Navajos, this tribe's distinctive language is derived from the Southern Athapaskan linguistic group. Today, the Athapaskan languages of these Diné people are also spoken by natives of northwestern Canada and Alaska. Apache mythology suggests a northern origin, but their customs and knowledge show a deep appreciation of southwestern resources.

The Apache tribe is divided among numerous bands, clans, and groupings that have separated, reunited, been conquered, or amalgamated. Their history is too labyrinthine to be summarized here, and most of it takes place outside this guide's range. Apaches have traditionally extended from northern Mexico and Arizona to the Great Plains as far as the Platte River in historic times.

As with other southwestern tribes, before acquiring the Spanish horse, the western Great Plains Apache bands hunted bison and other large game and moved from place to place using pack dogs. With the horse, as expected, they greatly expanded their ranges. Although many of the bands were mobile and lived by hunting, gathering, and raiding, others occupied river headwaters and cultivated corn, beans, pumpkins, and other crops.

In this guide's area, the Penxayes Apache band maintained camps in the Purgatoire Valley, probably in the early 18th century. Archeological evidence of their camps may have been found in the vicinity of the dam just west of Trinidad along Route 12 (see Chapter 7, Drives). Apparently there were no Apache bands active in this region during the mid-19th century period of American settlement.

PRE-SETTLEMENT TRAIL USED BY EUROPEANS

Spanish explorers in the Trinidad area in the 16th century often used trails that had been used by American Indians for centuries. Both French and American trappers and fur traders searched for beaver pelts to trade with the region's American Indians through the 1840s. Established trade routes connected Taos, New Mexico, with Bent's Old Fort near present-day La Junta, Colorado, which provided connection to St. Louis, Missouri. Direct routes to Denver were opened later.

In 1806, President Jefferson sent an expedition west led by Zebulon Pike whose task was to chart the southern boundary of the Louisiana Purchase. After a failed attempt to climb Pike's Peak, the expedition continued up the Arkansas River. From the Royal Gorge they traveled south down the Wet Mountain Valley and crossed the Sangre de Cristos, emerging at the Sand Dunes on January 28, 1807. It is uncertain whether they crossed Medano or Mosca Pass. Although hundreds of Americans had already entered present-day Huerfano County by that time, the Pike expedition marked the first official visit.

Trade routes crossed Huerfano County but led to no important destinations within the county until towns began to be established in the 1850s. The principal route on the north side of the Spanish Peaks was the Taos Trail, also called the Trappers Trail or Sangre de Cristo Trail (see Chapter 7, Drives, South Oak Canyon). It appears first in Juan Bautista de Anza's diary of his Comanche campaign of 1779 and was described vividly by Lewis H. Garrard in 1847. It is considered a branch of the Santa Fe Trail.

Passers-through included two well-known expeditions that were looking for railroad routes to the Pacific. In the summer of 1853, the ill-fated Captain John Gunnison took his government-sponsored expedition up the Huerfano River, then traveled southwest up the Taos Trail and over Sangre de Cristo Pass into the San Luis Valley. Gunnison was killed by a Ute band in Utah before the year was out.

In total disregard for Rocky Mountain winters, Colonel John Frémont led his privately funded fifth expedition up the Huerfano River in December of 1853. Solomon Carvalho, the expedition's daguerreotype photographer, said in his journal "...we followed that river to the Huerfano Valley, which is by far the most romantic and beautiful country I ever beheld." Traveling up Wet Mountain Valley the expedition crossed Medano Pass into the San Luis Valley, eventually camping near present Crestone. Later, the whole expedition was nearly lost because of severe weather it encountered.

TAOS TRAIL

Traveling from the north along the mountain front, the Taos Trail was a more direct, although less level, route to Taos than traveling south first to Santa Fe via Glorieta Pass. Beginning in Taos, the trail crossed Huerfano County at Sangre de Cristo Pass and proceeded northeast down South Oak Canyon to the Huerfano River (see Chapter 7, Drives) until it met the Mountain Route of the Old Santa Fe Trail north of Walsenburg. It then proceeded northward toward Pueblo and Laramie, Wyoming. This trail and others in the region were used constantly during the 19th-century trade period. Among those who probably used the route were the trappers Jules DeMun and Auguste Chouteau, Kit Carson, Lucien Maxwell, and Jim Webb. Few described it in detail, but it was certainly the most feasible route to Taos from Pueblo via the Sangre de Cristo Pass (near the present alignment of Route 160). In earlier times, it had been used by Utes who regularly crossed the mountains at this point.

SANTE FE TRAIL

Early in the 19th century, Santa Fe, then part of Mexico, was prevented from manufacturing goods or engaging in trade with the United States. The Spaniards wished to control all access to goods, even though citizens of Santa Fe were eager to trade. Two events in 1821 changed this. First, Mexico successfully revolted against Spanish rule, thus ending trade sanctions. Second, William Becknell, a trader from Missouri, pioneered a route west from Kansas City, then up the Purgatoire River and across a pass to arrive in Santa Fe. This route, or routes, later called the Santa Fe Trail, became a very important stimulus to commerce and settlement for what later became our southwestern territory.

The first route, called the Mountain Route, ran from Dodge City, Kansas, to Bent's Old Fort and crossed the Arkansas River at La Junta, and from there southwest in a nearly straight line to Trinidad, approximately along the alignment of Route 350. From there it ran south over Raton Pass into New Mexico, following the mountain front to the 7500-foot Glorieta Pass at the southern tip of the Sangre de Cristo Range. From there it was downhill into Santa Fe.

Initially, traveling with goods on the Santa Fe Trail was very risky, but by mid-century it was protected by a series of forts. The first, a commercial fort, now called Bent's Old Fort, was finished in 1834 by William and Charles Bent and Ceran St. Vrain. Army forts followed, including Fort Larned, Fort Lyon, Fort Union, and Fort Mann. After General Kearney took federal troops over Glorieta Pass in 1846 during the Mexican War, the Santa Fe Trail became a protected route over its length between Missouri and Santa Fe.

OTHER TRAILS

Raton Pass was a rough passage for wagons and therefore a major impediment for some travelers. The U.S. Army worked on it during the Mexican War in the 1840s. In 1865, Richens Lacy Wootton and George C. McBride improved it and ran it as a toll road but charged enough as to discourage many travelers. In 1867, Charles Goodnight balked at the exorbitant tolls and pioneered a new cattle trail over Trinchera Pass. The less circuitous Cimarron Cutoff of the Santa Fe Trail avoided toll, Trinidad, and the pass, and removed about 60 miles or 10 days from the trip. It traversed the corners of Kansas, Oklahoma, and northeastern New Mexico. Even though it lacked water, the time saved attracted many traders and settlers.

Finally, the nearest important nearby route to the west was the Old Spanish Trail that ran up the San Luis Valley from Santa Fe through Fort Garland to Saguache and points west.

EUROPEAN SETTLEMENT

In the 18th century, the present Huerfano County landscape was involved in a dispute between the French and Spanish, mostly due to differences in how the two cultures defined territorial boundaries. The Spanish governors claimed all of the land north of Mexico to a boundary represented by the Arkansas River itself. LaSalle claimed the entire Mississippi River watershed for France. In other words, the French used drainage divides or ridges to define territorial boundaries.

In the mid-18th century, various plains and mountain tribes seemed to accelerate raiding, fighting, and horse stealing. By the late 18th century, the Spanish governors of New Mexico saw their villages and settlements faced with major American Indian threats to their survival. (American Indians have a different perspective!) A succession of New Mexican governors had tried to defeat the Comanches, whose territorial center was at the eastern base of the Wet Mountains at the margin of the Great Plains. Comanche bands were successfully destroying settlements and taking and torturing prisoners.

To put an end to this, Spanish forces tried mostly frontal assaults, traveling east of Taos, then north along the mountain front toward the Arkansas River. Naturally, Comanche lookouts spotted the Spanish troop columns at great distance. The tribe always evaded or defeated the Spanish. During this period, the Comanches were led by a brilliant military strategist, Chief Greenhorn. The chief's name (in Spanish, *Cuerno Verde*) came from the striking horned headdress that he wore. It had belonged to his father, who was also a great chief in his day.

THE LAST CAMPAIGN: GOVERNOR ANZA AND THE COMANCHES

In 1777, Juan Bautista de Anza was appointed governor and military commander of New Mexico. Anza's military career in western Mexico had been distinguished. Among other accomplishments, he had led an expedition northward to California, where he founded the city of San Francisco in 1776, an event overshadowed by the American Revolution. As governor, Anza immediately saw to the proper fortification of his region, including Taos, against Comanche raids. He and his superiors doubtless were mindful that Spanish settlers had been driven back to Mexico by the Pueblo Revolt of 1680 and were loath to allow a reprise.

The summer of 1779 was to be Greenhorn's last chance to achieve his goal of expelling invaders from Comanche ancestral lands. Vengeance was also a factor, as Greenhorn's father had been killed by the Spanish. Anza's father, a military officer, had been killed by American Indians in Sonora, Mexico, when Anza was a young boy. When Anza was made governor, his mandate was to protect Spanish settlements at whatever cost. So all elements were in place for a classic last stand. Two iron wills and two brilliant military strategists, each man was driven to settle the score and complete the task set before him.

That summer, Anza organized a military expedition with the goal of destroying Greenhorn and the Comanche will to fight. With 600 troops, 1800 horses, and 200 revenge-bent Utes and Apaches, Anza spent most of August 1779 traveling a widely circuitous route, intent on surprising Greenhorn's warriors from the north. From Santa Fe, Anza's column moved up the San Luis Valley to its northern apex, turning eastward near present-day Salida, Colorado. Anza crossed the South Platte River, moved on to Florissant, and crossed Ute Pass. Near the end of August, his force had moved south to Colorado Springs. Historians regard this as a truly audacious strategy for an 18th-century military campaign. The force was large for a mountain wilderness, considering the absence of roads, and trails were often vague.

On September 1, at the edge of the plains between Colorado Springs and Pueblo, Anza engaged the Comanches with almost complete surprise. Eighteen warriors were killed, 60 women and children were taken prisoner, and 500 horses were taken. The next day, Greenhorn's warriors planned an attack on the Spanish camp, which was hastily aborted when they discovered that Anza had troop columns arrayed to surround them. Early on September 3, Anza advanced southward to what became Greenhorn's final field of battle.

Again, Anza had divided his forces into several parallel columns. Partly due to the chief's bad luck, Anza succeeded in cutting off Greenhorn's party, which included his son and heir apparent, his top chiefs, and the medicine man who had convinced Greenhorn and his followers that Greenhorn was immortal. All died in a ravine from which there was no escape.

The precise location of the battle was once believed to be at present-day Greenhorn Park, just west of Colorado City and just south of Route 165 (Fig. 7). However, Martinez (2001) argues convincingly that the conflict took place 9 miles northward at a gully of the St. Charles River* (Fig. 8). Wherever this historic battle's exact site, the century-old Comanche dominance of the plains was finished.

As best as can be interpreted from his unusually detailed diary, Anza's military expedition made remarkably good time when underway. They returned toward Santa Fe the same day of the battle, camping along the Huerfano River, perhaps near the historic townsite of Badito (see Chapter 7, Drives). The following day, the expedition cut diagonally across Huerfano County, probably following the "Old Taos Trail," and went over Sangre de Cristo Pass (Fig. 9). This low ridge is just north and west of present North La Veta Pass, now tra-

*The suggested site is at the junction of Water Barrel and Burnt Mill roads at the bridge over the St. Charles River. Take I-25 to exit 88, west on 247-246 roads. Left on Road 230 to Burnt Mill at junction of Road 221, or Water Barrel Road.

Fig. 7: Cuerno Verde marker in a park just west of Colorado City and just south of Route 165. This marker commemorates the life of the great Comanche chief who was killed in battle by Governor Anza's forces in 1779. This park may not be the actual site of his final battle (see text discussion).

Fig. 8: The St. Charles River Valley near the intersection of Water Barrel and Burnt Mill roads, which, according to Martinez, is probably where Chief Greenhorn fought his final battle.

Fig. 9: South Oak Canyon, looking southwest toward Sangre de Cristo Pass. At present, this is private property. It was the probable route of Anza's forces on their return to the San Luis Valley.

versed by Highway 160. The army camped that night near Fort Garland. (See Chapter 7, Drives, South Oak Canyon, for a way to view their route.) This may be the first historically recorded use of the Taos Trail.*

Years later, in February 1786, a peace treaty formulated by Colonel Anza and the new Comanche Chief Ecueracapa was signed. Anza delayed the signing until all Comanche bands agreed to sign.

Occasional clashes with American Indians continued some decades into the future. The Utes had had only occasional contact with whites who ventured into their domain. However, after 1848 they felt the full impact of white civilization. Territorial loss was followed by defeat and disillusionment. By the mid-19th century, what later became the American Southwest was considered safe for settlement.

Fittingly, Cuerno Verde is immortalized by Greenhorn Peak, which forms the highest summit of the massive Wet Mountains that loom over the Huerfano

*The trail became the principal route between Pueblo and Taos for more than a century. Travelers going south from the Pueblo area turned westward at Huerfano Butte, followed the Huerfano River, and entered the mountains. From 1821 to early 1850s, the Taos Trail was used largely by mountain men going to Taos for trade and to marry Mexican women.

River Valley to the south and Comanche territory to the east. The peak can be seen from nearly any point in the county. To the south, Anza is commemorated by De Anza Peak, a 13,362-foot feature in the Culebra Range. It is directly west of Monument Lake on Route 12 and just over the Costilla County line.

THE LAND GRANTS

In 1821, Mexico gained independence from Spain. An immediate problem for the Mexican government was proving its claim to the southwestern United States up to the Arkansas River on the northeast. From the late 1830s through the early 1840s, the Spanish governors of New Mexico, from their capital at Santa Fe, were authorized to issue land grants extending as far north as central Colorado. The grants were intended to hasten settlement of that vast expanse by Spanish settlers, thereby securing the territory and its resources for the Spanish way of life. It was hoped that settlement and cultivation would fend off hostile American Indian and American settlers, as well as legitimize Mexican claim to all territory up to the Arkansas River. As a condition of their land grants, grantees were pledged to defend Mexico against American westward expansion. At that time, these lands were inhabited almost entirely by native people, and available Mexican troops were inadequate to secure the peace.

VIGIL-ST. VRAIN LAND GRANT

At more than 4 million acres, the Vigil-St. Vrain tract covered nearly all of the area of this guide and more. It included the watersheds of the Huerfano, Cuchara, and Las Animas (Purgatoire) rivers, from the crest of the Sangre de Cristos, to near the confluence of the Animas River at the Arkansas. If it had remained intact, it would have been the world's largest tract of private property. Obtained by Cornelio Vigil and Ceran St. Vrain in 1844, the owners did little development aside from convincing some ranchers to graze cattle. Instead, they sold off chunks to Charles Bent, Governor Armijo, who issued the grant in the first place, and others. In turn, before 1846, heirs sold off much of their land in anticipation that the grant scheme would be overturned.

The neighboring Maxwell Land Grant to the south included about 2 million acres, most of Colfax County, New Mexico, and was the largest in New Mexico. To the west, the Sangre de Cristo Land Grant covered 1 million acres of the eastern San Luis Valley. In terms of encouraging settlement, the Sangre de Cristo grant was the most successful because it was on a level and direct supply route north from Taos and Santa Fe. Many of today's eastern San Luis Valley property owners can trace their deeds to the original grant.

Through 1845, the Spanish governors of New Mexico continued settlement efforts to solidify claims to southern Colorado. East of the Sangre de Cristos, however, settlement lagged. Huerfano and Las Animas counties were too remote from supply routes or garrisons to protect settlers.

THE AMERICAN PERIOD

By the mid-19th century, settlement pressure and land ownership ambitions by the United States intensified. In 1845, President Polk annexed Texas, thus precipitating the Mexican War of 1846. After a number of battles, U.S. troops under General Winfield Scott finally defeated General Santa Anna's Mexican forces, which led to the Treaty of Guadalupe Hidalgo in 1848. The greatly weakened Mexican government ceded all claims to California and lands south of the Louisiana Purchase, which added about 1.2 million square miles of southwestern land to the United States. Although the land grant schemes ceased to exist, property rights were confirmed for some Mexican settlers on some grants, including the Vigil and St. Vrain Land Grant. Until 1891, the U.S. Court of Private Claims worked to sort out valid land claims on a case-by-case basis. The problem for most hopeful "owners" was that few could afford surveys or otherwise define their claims.

In the 1860s, especially following the end of the American Civil War, Anglos arrived from the east bringing a very different, individualistic settlement pattern. The "little house on the prairie" best describes their approach. Settlers worked their way south from Denver, an early hardscrabble jumping-off point in those days. Army posts provided protection and ranchers supplied beef to the army posts. In Huerfano County, the Huerfano River Valley became most densely settled due to its broader and more cultivatable acreage than the narrower Cuchara Valley. Gardner is the sole surviving town among the many settlements that once existed.

HUERFANO COUNTY

A conical volcanic plug at the edge of the Great Plains was called *El Huerfano*, "the Orphan," by Spanish explorers in 1806. Visible from a distance by travelers crossing the region, it became a landmark along the Mountain Route of the Santa Fe Trail. It arises near the crossing of I-25 and the Huerfano River, about 10 miles north of Walsenburg.

By 1861, the U.S. Congress established Colorado Territory, and soon thereafter its first assembly created Huerfano County among 17 others. Larger than many eastern states, Huerfano County was the largest of these early counties. Its boundaries extended north to the Arkansas River, west to the crest of the Sangre de Cristos, south to the New Mexico border, and east to Kansas. Its first county seat was Charlie Autobees' Plaza, a population concentration of a few families at the junction of the Huerfano and Arkansas rivers, east of present-day Pueblo. With only a few trails and fewer than 100 people in this vast region, little of the usual county business could be conducted. Five years later, after the Civil War (1866), new counties, including Las Animas County, were sliced off on the north, east, and south. Greatly reduced in size, Huerfano County was provided an equal division of mountains and plains.

In 1866, Badito became the first county seat to be located within Huerfano's present boundaries. This small community (see Chapter 7, Drives) had been settled in the late 1850s and was located on the Taos Trail at a ford over the Huerfano River. George Simpson, who lived in Badito, was appointed first county clerk by the territorial governor. In those days, the county seat, by law, was where the clerk lived. The site's pleasant position, where the Wet Mountains reach the plains, proved too isolated from centers of county growth and development to the east. In 1872, the more populated Walsenburg became county seat and remains so. Today, crumbling ruins and dignified old cottonwoods mark Badito's site along the Huerfano River 9.3 miles east of Gardner and just north of Route 69.

Emerging Population Centers

In 1862, abundant game was noted in the region, including bison, antelope, deer, elk, and sheep. Black and grizzly bears and mountain lions were common in the hills, as were beaver and muskrat. After the Civil War, settlers arrived from the east in ever-increasing numbers, and, within the next two decades, most large animals were extirpated. Several towns sprang into being along the Cuchara River between La Veta and Walsenburg, but only those two towns survive.

The site where the Cuchara River emerged onto the plains proved to be centrally located for travelers arriving from all directions. First known as Plaza de Los Leones, it was named for the family of Don Miguel Leon, who settled there around 1859. Soon, various entrepreneurs followed, such as Fred Walsen and the Sporleder and Unfug families. Walsen was a natural leader who renamed the town Walsenburg and established its first post office in 1870.

Incorporation followed in 1873 with Walsen as first mayor. By 1874 the population had risen to 150. When looking at the creek-like Cuchara River today, it is hard to imagine the catastrophic floods that occurred in 1878 and 1910 and caused considerable damage in Walsenburg.

Commerce

In the 1870s, businesses and other development accelerated. By 1876, the Denver and Rio Grande Railroad reached a temporary terminus in La Veta while the company surveyed for a route into the San Luis Valley. Soon after, former governor A.C. Hunt, the company's managing director, finally found a useable route. A sharp hairpin turn that became known as "Hunt's Goose Egg," or "Muleshoe Curve," was considered nearly impossible by Hunt's engineers, but it was successfully built. In June 1877, the first passenger train went over La Veta Pass at 9339-foot elevation. It was known for a time as the highest railroad pass in the country. The narrow-gauge trains operated over this pass until 1899, when the route was moved to a lower-elevation pass 7 miles to the south. Today, a road following the original railroad alignment over Old La Veta Pass makes a loop beginning and ending on Route 160 (see Chapter 7, Drives).

The second track alignment, in standard gauge, followed Middle Creek, a tributary stream of the Cuchara River, then climbed over the 9240-foot Veta Pass. These trains began running in 1900. In addition to freight, passenger service continued until 1953. After a 53-year hiatus, in 2006, the Rio Grande Scenic Railroad established the San Luis Express, an excursion train running between La Veta and Alamosa (see www.alamosatrain.com). It is advertised as the highest standard-gauge pass over the Front Range.

As soon as the original train transportation was established, the coal that was long known to exist in the area could finally be exploited and sent to market. Fred Walsen's Walsen railroad line, 1 mile west of Walsenburg, led to development of the mining towns of Cameron and Robinson.

LA VETA

Founded in 1866, the town was originally known for its founder as Francisco's Ranch or Plaza. The origin of the modern name is uncertain. In Spanish, *la veta* means "the vein." Hiram Vasquez (1843–1939) said the town was named by Mexicans for a vein of white mineral; "La Veta Tierra Blanca" was located at the base of Piñon Hill at that time. Also called "yaso," the mineral was used by settlers to whitewash their dwellings. The problem is that no such candidate vein is known. Another theory is that La Veta is a corruption of Abeyta, the name of a Hispanic family who lived in the La Veta Pass area.

"Colonel" John M. Francisco had been a trader between St. Louis and Santa Fe after he explored western routes beginning in 1839. Located in Taos until 1851, he moved to Fort Massachusetts, the forerunner of Fort Garland in the San Luis Valley. He was an army sutler (storekeeper) there until 1862, and his title "Colonel" was honorary. Francisco's La Veta interest was based on his purchase of thousands of acres of land on this future townsite from the Vigil-St. Vrain Land Grant. He had selected the site for a home while at Fort Garland.

Francisco formed a business partnership with Henry Daigre (1832–1902). They farmed and built their plaza or fort on the main trail leading from the Mountain Branch of the Santa Fe Trail to the San Luis Valley. With their other partners, Hiram Vasquez and John Delivera, Francisco and Daigre traded with Americans, Mexicans, and American Indians and did some farming. Vasquez had led a colorful life. His stepfather was a partner with Jim Bridger in Indian trading in Utah and Wyoming. Over time, Francisco and his partners sold off lots to settlers. Even though later title disputes caused Francisco to lose most of this wealth, he still led an active community life.

The north and west portions of the adobe plaza built by Francisco in 1862 still stand in the center of La Veta and are now the core of a well-organized historical museum. A huge cottonwood, said to have been planted by Francisco in 1878, stands within the plaza. Typical of plaza towns of that era, Francisco's plaza was originally a square, 100 feet per side, surrounding a well. It had a wide entrance on the north side that could be closed by a heavy gate. The enclosure was large enough to shelter wagons and stock against American Indian attack. The one-story buildings had thick dirt roofs that sloped inward and were surmounted by a parapet that protected riflemen.

In the last quarter of the 19th century, development accelerated as was common among western settlements on transportation routes. Cattle ranching and agriculture prospered, and the first barbed wire fence was built in the late 1860s. Churches were erected in the 1870s and 1880s, along with the first school and the first post office, which was within Francisco Plaza.

Fortunately for La Veta, the Denver and Rio Grande Railroad (D&RG) arrived in 1876. Trains stopped at the forwarding station inside the plaza. Later a railroad station was built a few blocks north, near the present La Veta town park. During the years 1876 to 1882, La Veta was governed by a board of trustees whose first chairman was Francisco. By 1882, American Indians were gone from the region and the town thrived.

THE GOLD RUSH

The Colorado Gold Rush of 1858 near present-day Denver greatly swelled its population. It occurred to some that the Spanish Peaks might be a good place to find gold (or silver), so in the 1870s about 50 to 60 shafts were dug or blasted. Robert Lincoln, son of President Lincoln, had shares in the Bulls Eye Mine on West Spanish Peak. The tiny amount of gold or silver found in these mines seldom justified expenses.

DEVELOPMENT ALONG THE HUERFANO RIVER

Compared with the Cuchara Valley, the broader Huerfano River Valley is ideally suited for agriculture, and town expansion never developed momentum. Today, one-time towns such as Badito, St. Mary's, and Malachite appear only on maps. Agriculture remains the central pursuit of the valley (Fig. 10).

Gardner

The town, which was named for Herbert Gardner, an early settler, is the last survivor among the original settlements. Never incorporated, it remains a stable settlement in the center of a rich agricultural region. Its early history

Fig. 10: Historic stone school building at Malachite.

remains obscure. Once home to one of the largest regional stores, complete with hitching post, today it seems too isolated to attract a viable commercial presence.

Farisita

This townsite just west of Badito was a small cluster of homes, but perhaps two remain. A regional trading center once, it had a general store and post office. Infamous western desperados reputedly passed through the town. It was named in Spanish for Jeannette, "the little Faris girl," daughter of first postmaster, John Faris.

LAS ANIMAS COUNTY

The county began its identity in 1866 having been carved out of the huge original Huerfano County to the north. In 1542, Coronado's expedition touched this part of southern Colorado, but the river arising from the Sangre de Cristos south of the Spanish Peaks remained unnamed for the next 50 years.

The Purgatory or Purgatoire River

The county is named for the principal river that flows through it, and the following legend explains the original gloomy name: *El Rio de Las Animas Perdidas en Purgatorio* (The River of Lost Souls in Purgatory). Around 1593 to 1595, the unauthorized Bonilla and Humaña Expedition headed northward out of Mexico, allegedly to suppress American Indians and then seek fortunes on the side. Along what is now known as the upper Purgatoire River Valley, the leaders quarreled and Bonilla was slain.

The expedition priest refused to give last rites because of the illegality of the expedition. Some members had already returned to Mexico as the whole expedition had been so ordered. Humaña and the rest continued on until one morning, when their camp was burned by American Indians and most members died on the banks of the river. Many years later, the expedition's rusting Spanish arms were discovered and it was realized that the members had received no last rites. The long name thus memorializes this event.

Over the last hundred years, the names Las Animas, Purgatory, or Purgatoire (a translation offered by French trappers) can be found on local maps as place names, as well as names for the river itself. The name "Picketwire," now a place name, was probably corrupted from Purgatoire by Anglo cowboys. In 2005, a local petition was submitted to the U.S. Board of Geographic Names to reinstate the original long form, but it failed approval.

Trinidad

Trinidad is the largest settlement in the county, and it conveniently anchors the southeastern corner of this guide's territory. The town was well placed astride the Mountain Route of the Santa Fe Trail where it crosses the Purgatoire River. Its early growth and fame were based on its position as the first major town that travelers encountered after many days on the Great Plains. Today, it has a nice historical district that commemorates this early history. (See Chapter 7, Drives).

In the 1850s, small numbers of Hispanics, who settled in family groups, arrived from the south. Some entered the upper Purgatoire Valley to the west, where a string of towns began as traditional plazas. Such towns can be seen today in many rural precincts of Latin America. Most plazas had a central well. When under siege, plazas could be closed off in defense of settlers and their animals. By the late 1860s, Trinidad had grown to about 1200 people living in mostly one-story log or adobe houses. The first substantial two-story home was built in 1870 on Main Street, the old trail route, by trail merchant John Hough. His family later sold the property to the Baca family, and the house is now part of the Trinidad Historical Museum. It features a rather complete collection of family and period furnishings.

In 1876, the year Colorado became a state, Trinidad was incorporated. Two years later, the Atchison, Topeka and Santa Fe Railroad (AT&SF) arrived in town and then placed tracks over Raton Pass the next year. Tracks to Santa Fe were completed by 1880. The railroad surely accelerated the region's prosperity even as trail trade declined. In the last quarter of the century, cattle ranching prospered along with diverse agricultural commodities.

A major stimulus to the economy was coal extraction, a resource long known to exist but which awaited a means of transport. Coal camps were established throughout the Purgatoire Valley and along the mountain front all the way to the north of Walsenburg in central Huerfano County (see Chapter 4, Geology, for names of the coal-bearing strata). Some had schools and recreation halls, otherwise rare in the region. At the turn of the century and up through World War I, thousands of foreign immigrants arrived in southern Colorado, where most lived and worked under dismal conditions. The United Mine Workers

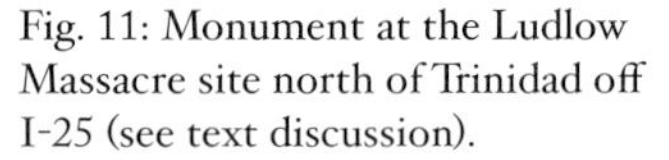
Fig. 11: Monument at the Ludlow Massacre site north of Trinidad off I-25 (see text discussion).

attempted to organize the miners while the companies established a militia to thwart these efforts. The strikes of 1913–1914 were major events in the history of American labor relations (see Chapter 7, Drives, Ludlow Massacre site) (Fig. 11). Unfortunately, there were numerous confrontations and finally many deaths at Ludlow in April 1914 as the militia sprayed the tent camp with machine guns and then burned it down. More deaths of miners followed at other coal camps as well (see McGovern and Guttridge [1996] for definitive coverage).

An unfortunate geographic feature of this modern town of more than 13,000 people is that its nice riverbank position in the Purgatoire Valley has been additionally gashed, first by railroad tracks and then by the elevated I-25. But it is easy to get past this first impression by exploring the town from its streets and sidewalks. Here, you will find a fascinating southern Colorado gem that proudly preserves its heritage (see Chapter 7, Drives).

CHAPTER 4

REGIONAL GEOLOGY: A BRIEF GUIDE

COAUTHORED BY GORDON N. KEATING

THIS COMPACT region reveals a remarkable array of landforms produced by a complex and tortured history. The following description, jargon-laden in places, is followed by a glossary at the end that we hope will be helpful. We are indebted to numerous authors, especially Aber, Baldridge, and Johnson, cited in Resources. A field trip in the region led by David Moore of the U.S. Geological Survey was very instructive.

THE SETTING

Our striking physical landscape is due to events and processes that depend on time scales ranging from thousands to billions of years. The processes that change our earth today are much the same as those that occurred in the distant past. The big picture of today's geologic narratives is based on the plate tectonics revolution that swept the field in the 1960s. Many of the earth's surface features are clues in the story of a dynamic earth where a thin and brittle crust continues its rearrangement over a hot, viscous mantle approximately 22 miles down. The crust itself is fragmented into continents and lesser segments that are called *plates*. The crustal plates, including the adjacent upper mantle, remain in constant motion and move at roughly the rate that fingernails grow. Their most stable and thickest parts are cratons, such as the Canadian Shield. Adjacent thinner areas, such as the Spanish Peaks region, are called *platforms*.

Plates change shape, too. A billion years ago, North America and other continents would not have been recognizable. Since that time, the earth's plates have collided, sometimes gathered together into supercontinents, then separated or were subducted (overrun and remelted).

The Spanish Peaks, at the edge of the Great Plains, are at the platform margin of the North American craton. West of our region, about 800 miles of land, also called *exotic terrane*, has been added by the collision or subduction of adjacent plates originally off the North American coast. This occurred roughly between 350 and 40 Ma.* Collision pressure causes periods of compression and crustal thickening, regional uplift, and orogeny (mountain-building events). From the Spanish Peaks westward, there have been multiple cycles of crustal folding, thrust faulting (where strata override one another), and uplift, all alternating with episodes of erosion to base level.

Later, when released from such pressures, the land extended and the crust thinned. Extension was accompanied by normal faulting, which led to formation of deep valleys between ranges. Mountain building and the spread of surface magma can occur in either compression or extension, based on crustal instability within or at the edge of a platform.

TIME AND CHANGE: EVENTS OF THE LAST HALF-BILLION YEARS

THE OLDEST ROCKS

(See Table 1 for an outline of geologic time.) Although the earth was formed about 4.54 Ga, the oldest exposed Precambrian rocks in the Southwest are Proterozoic metamorphic rocks formed about 1.7 Ga. The original ages are hard to decipher because they were probably recycled (remelted) from yet older material. They are exposed in places as the cores of the Sangre de Cristo and Wet Mountains after alteration by heat, pressure, and time at depths of about 6 to 12 miles. Precambrian rocks are also exposed at a road cut on the south side of Route 160. Descending from La Veta Pass into the San Luis Valley, one passes the small ghost town (two buildings) of Russell on the right. On the left side are rocks formed 1.7 Ga that are cut by a slightly younger dike composed substantially of hornblende.

The geologic eras and periods within them are defined by relatively abrupt changes in the fossil record or sequences of deposits (strata). A given event or stratum can be timed, relatively, by its relationship to rocks above and below. Absolute age can be worked out by analyzing radioactive decay of elemental isotopes, a highly technical specialty.

*For time periods, we use *Ma* to represent million of years ago; *Ga*, billions of years ago; and *Ka*, thousands of years ago.

Table 1. Geologic Column for Spanish Peaks Region

Series	Ma	Formation	Thickness	Texture, Notes
Holocene	0.01	Recent	10–20 feet	Alluvium, talus, loess
Quaternary	1.8	Pleistocene		Gravel, sand, silt Terraces, moraines, alluvium
Oligocene	33.7			Igneous intrusives, eruptives (e.g., Spanish Peaks)
		Huerfano	0–2000 feet	Reddish cobble and boulder conglomerate
Eocene	54	Cuchara	0–5000 feet	250–350 feet in Raton Basin
Paleocene	65	Poison Canyon	0–2500 feet	Yellowish, arkosic, and conglomerate beds that weather irregularly; no coal
		Raton	0–2075 feet	Shale, sandstones; conglomeratic or coal bearing
Upper Cretaceous	90	Vermejo	0–360 feet	Coal bearing
		Trinidad	0–255 feet	Sandstone
		Pierre	1300–2900 feet	Shale
		Niobrara	400–500 feet	Marls or limestone
	95	Greenhorn	100 feet	Marls or limestone
Lower Cretaceous	146	Dakota	100–200 feet	Thick-bedded quartz sandstone and gray shale; usually deposited directly on Morrison
		Purgatoire	100–150 feet	
Upper Jurassic	160	Morrison	150–400 feet	Red and greenish shales
Middle Jurassic		Wanakah	?	Sandstone and limestone
		Entrada	?	Buff sandstone, red siltstone
Permian/ Pennsylvanian	286	Sangre de Cristo	Thousands of feet	Continental sediments, red beds, not fossiliferous
Mississippian		Pass Creek	?	Sandstone, non-marine
Precambrian	1700		?	Metamorphic, crystalline

Sources include: Johnson, R. B. & J. G. Stephens. Geology of the La Veta Area, Huerfano County, Colorado. USGS Oil and Gas Inventory, Map OM-146, 1954; Geology of the Huerfano Park area, Huerfano and Custer counties, Colorado, Geol. Soc. Amer. Bull. 63(9): 953–992, 1960; and Worrall, J. Preliminary geology of the Oakdale Field, Northwest Raton Basin, Huerfano County, Colorado. American Association of Petroleum Geologists Southwest Section Convention, Ruidoso, New Mexico, 2002 (http://www.searchanddiscovery.net/documents/2004/Worrall).

THE PALEOZOIC ERA (543 to 251 Ma)

The first significant fossils appeared in sedimentary rocks of Cambrian time, and this onset of conspicuous evidence of life is what defines the era. Since then, almost continuously, extinctions and species changes among fossils in ever-younger rocks have allowed geologists to identify and arrange strata in datable sequences and thus interpret the earth's history.

Throughout the Paleozoic, this landscape alternated between terrestrial and marine. Marine transgressions over the landscape lasted millions of years, and adjacent elevated land surfaces were the source of sediments that eroded and accumulated as thick sedimentary rock, a process that occurs on oceanic and lake bottoms today. The long sedimentary record of these eras is best seen in canyons or road cuts, or on the sides of cliffs.

First Mountain-building Event: The Colorado Orogeny

In the Late Devonian period (417 to 354 Ma), we have the first evidence of an active western margin of the North American Plate. The collision of plates or island arcs south and west of here added thousands of square miles of real estate to North America. The resulting pressure at the North American margin began in the Late Mississippian period (354 to 323 Ma), reached its greatest intensity during the Pennsylvanian, and persisted into the Permian (290 to 251 Ma). Compression and deformation resulted in folding and thrust faulting. About 80 to 100 miles of continental shortening took place accompanied by the elevation of a series of high ranges bounded by deep basins. These ranges, known as the Ancestral Rocky Mountains, or Colorado Orogeny, included southern Colorado and adjacent New Mexico.*

During the Permian, as the Ancestral Rockies rapidly eroded, sea levels declined and marine sediments were overlain by continental deposits. At the same time, an arid climate led to the formation of great dunes and red beds across wide areas including the Colorado Plateau to the west. Today, the ancient Rockies are recognized as the source, through erosion, of great fans of continental sediments thousands of feet thick that nearly filled adjacent basins. A long time into the future they were to be uplifted again by similarly powerful forces.

*Also at this time, North America collided with Africa, which resulted in the elevation of the Appalachian Mountains in the eastern United States and the Atlas Mountains in northwest Africa.

A Great Mass Extinction

The end of the Permian period, and thus the Paleozoic era, is marked by the greatest mass extinction event in the history of the earth. The cause is yet unproven. Candidates include an asteroid impact, supernova radiation, and volcanism. At that time, there was a million-year-long outpouring of 1.5 million cubic miles of basalt lava onto the Siberian landscape. Analyses of geologic strata of that period show that the event was ecologically catastrophic. Ninety-six percent of marine species perished, as did 70% of land animals. Most terrestrial plant populations essentially disappeared for the next 10 million years. Likewise, reef building in shallow seas ceased completely. Strata of that age record a sharp increase in remains of fungi, nature's scavengers. Carbon dioxide accumulated in the atmosphere to high levels, possibly as a result of a severe drop in photosynthesis.

THE MESOZOIC ERA (251 to 65 Ma)

During the Triassic period (251 to 206 Ma), seaways invaded the land, and much of the central and northern Rockies region was covered by a wide oceanic bay. By then, the edge of the North American platform was in today's central Utah and southern Nevada. Beyond this continental margin was a fluctuating shoreline that was at low latitude, tropical, humid, and tectonically quiet. There is little record of the great extinction recorded in these strata. The adjacent Southwest became more arid, as it was now in the trade wind belt, 13 to 16 degrees north of the equator. Deserts and dune fields covered the region.

In the Cretaceous period (144 to 65 Ma), western North America was invaded by the western interior seaway that was about 600 miles wide and 180 feet deep in Colorado. The sea advanced and withdrew during its five-million-year existence. Impressive sedimentation rates led to accumulations of more than 100 feet per million years in parts of the interior sea, especially from river deltas and marine margins. In this region, its legacy is the Upper Cretaceous Dakota Formation, a conspicuous sandstone outcrop. Dinosaur tracks are preserved in some parts of the formation.

The Collision That Roughened the Western Landscape

Northward drift of the North American Plate continued while the oceanic Farallon Plate struck the western margin of North America. Throughout the Cretaceous and into the Mid-Eocene, the huge and denser Farallon Plate was subducted beneath the lighter North American Plate until it was eventually consumed. The collision began as a perpendicular, high-angle force that elevated the Sierra Nevada and Baja California uplands and resulted in 75 miles of compression of the western landscape. The compression produced the Laramide orogeny, stretching all the way into southern Colorado. Before this orogeny began, much of the previous mountainous uplift had eroded to near sea level. Then, coincidental with renewed uplift, the Cretaceous period and the Mesozoic era ended in a catastrophic ecologic collapse.

A Hit from Space That Changed the World

Sixty-five million years ago the end to the Mesozoic era was cataclysmic. An asteroid struck the earth with a force of 100,000 megatons at Chicxulub ("chic-shu-lube") Crater on the northern Yucatán coast. Once again, mass extinctions and abrupt changes occurred in the geologic record. After the impact, it is calculated that atmospheric dust darkened the sky for years, causing the earth's average temperature to drop by 22°F to 34°F for up to a decade. Up to one fourth of the earth's biomass may have burned in the aftermath. Huge amounts of carbon and debris rained down for months. It was an unprecedented ecologic disaster, especially here and to the southwest (see Alvarez [1997] for a lucid description of this event).

Early evidence of this event was discovered near Raton, New Mexico, and Trinidad, Colorado, in boundary strata between the Cretaceous and Paleocene (see Raton Formation below). In many other parts of the world, within and adjacent to coal swamp sediments, fallout from the impact has been discovered. In the band of deposits resulting from aerial dust, one finds anomalous quantities of iridium (a platinum group element, common in meteorites but rare on Earth), shocked quartz, feldspar and zircon, glassy tektites, nickel-rich spinel, diamonds, sometimes elemental sulphur, carbon, soot, and fullerenes (ball-shaped carbon molecules). Judging by coal deposits that overlie this layer, recovery of the flora was more rapid in contrast to the impact that ended the Paleozoic era.

CENOZOIC ERA EVENTS

Second Mountain-building Event

The Laramide orogeny continued (70 to 35 Ma) through the impact period, causing the uplift of a second series of Rocky Mountains. Sediments accumulated in Mesozoic seas, forming a major part of the new mountains. As uplift continued, the mountains eroded at accelerated rates, often exposing Precambrian rocks. Mountains of the Front Range, the Sangre de Cristo Range, and ranges in New Mexico were elevated in the same places or about 50 miles east of the long-vanished Ancestral Rockies. It seems that crusts and the adjacent upper mantle, having once been cracked by an original orogeny, are vulnerable to future stresses.

From Late Cretaceous time onward, the region has remained above sea level, such that accumulation of sediments is exclusively from eroding terrestrial sources. When the Laramide orogeny ended in Late Eocene, it left the West with numerous ranges and a thickened crust with an average elevation of about 9500 feet.

The trigger or pressure causing the Laramide uplift was probably a change in the subduction of the Farallon Plate. The subduction, at an increasingly shallow angle, eventually became nearly horizontal. It affected the entire landscape from the edge of the Great Plains westward. Finally, between 40 and 28 Ma, the Farallon Plate was almost completely consumed (melted) beneath North America, and the Pacific Plate abutted the western continental margin. This time, rather than colliding perpendicularly, the Pacific Plate began sliding parallel to the boundary with the North American Plate and is marked today by the San Andreas strike-slip or transform fault.

But we are not yet done with mountain building (orogenies). The ranges we see today in the West are mostly not from the Laramide orogeny. Parts of the Front Range and southern Sangre de Cristos of New Mexico date from this period, as 40 million years of erosion was insufficient to remove them entirely. Only in the southern Rockies can one find the record of mountain-building events that span the entire Tertiary.

MID-TERTIARY TO PLEISTOCENE (35 to 2 Ma)

Third Mountain-building Event: The Modern Rockies

After 250 million years of compression, a change to continental extension led to a thinning of the crust. This process was accompanied by normal faulting and resulted in the uplift of long, narrow mountain ranges, such as the Sangre de Cristos, that are separated from adjacent ranges by broad, sediment-filled valleys. To the west, the Rio Grande rift is a prominent example. The rift began opening about 27 million years ago and is propagating northward.

During the Basin and Range extension, the West subsided to its present average lower elevation. Renewed magma formation occurred over the Southwest and locally. The third wave of mountain building continued due to instability and turbulence in the wake of material from the Farallon Plate sinking into the mantle. Just in the last 4 to 7 million years, the total uplift of the southern Rockies has averaged about 3500 feet.

During the reprise of the third mountain-building event (growth of the modern Rockies), the Colorado Plateau to the west and Basin and Range deserts assumed their modern appearance. Volcanic calderas and other magma formations dating from the period 34 to 22 Ma were established in central and northeast New Mexico, as well as here in southern Colorado. Extrusions of lava in the Raton Basin date from this time.

Locally, magma bodies (plutons) welled up into Tertiary sediments and cooled. The Spanish Peaks and the several other plutonic bodies in the region were exposed later by the erosion of the surrounding softer sedimentary layers. Additional unexposed metamorphic deposits also occurred in the region. The processes of uplift and erosion of the modern Rockies continue to the present.

PLEISTOCENE EPOCH TO RECENT (2 Ma to 10 Ka)

In the Northern Hemisphere, a cooling climate led to four major advances of glacial ice in the past two million years. As reviewed by Aber, there is much evidence of glaciation in the Culebra Range in the vicinity of the Spanish Peaks. On the high peaks, glaciers carved characteristic bowl-shaped cirques and rounded valleys. These can be seen at higher elevations on the Spanish Peaks, especially on the north side of West Peak (Figs. 12 and 13) and on the northeast side of Mount Trinchera, where there are several classic cirques.

All but the most recent ice advances are poorly represented in this region. From the last glacial advance, three glacial valleys originated on the east front of Trinchera Peak and descended the Cuchara Valley to a lower limit of about 10,000 feet. They date from Pinedale (Late Wisconsin times, 23 to 12 Ka). During the glacial retreat, the ice margin paused at about 10,400 feet where resistant bedrock of the Cuchara Valley constricted movement of the three glacial tributaries. This can be seen in the vicinity of Blue and Bear lakes, where a moraine complex consists of lateral and end moraines that impound these lakes. Glacial till of these moraines can be seen at road cuts in the vicinity and consists of large gray and red sandstone fragments in a finer matrix. Larger rocks bear the striations and grooves that mark glacial movement.

Fig. 12 (above): Looking south near the Bulls Eye Mine site on the north side of West Spanish Peak. A terminal moraine appears in the middle distance of this typical cirque basin.

Fig. 13 (left): V-cut in terminal moraine drained a cirque lake near the Bulls Eye Mine site. This may have been formed by natural rapid draining of the lake during the Holocene, or it may have been encouraged by miners.

Periglacial Features

On alpine ridges, postglacial features date from cold episodes during the Holocene. Intense freezing and thawing patterns, sometimes at depth, caused patterned ground and fell fields, or rock glaciers. On the valley floor east of La Veta, the outwash terraces north of the Spanish Peaks may have been left behind by rapid runoff events in the Late Pleistocene.

REGIONAL FEATURES

The geomorphology, or form of the landscape, is largely a result of recent processes of erosion and uplift. By recent, we mean mostly in the last several million years. Older rocks are exposed by erosion and uplift.

THE RATON BASIN

East of the Sangre de Cristo Range the highly dissected upland on the western margin of the Great Plains, including Huerfano and Las Animas counties, is the Raton Basin. This 2500-square-mile asymmetrical landscape includes south-central Colorado and north-central New Mexico. From the Great Plains (6000 feet) to the summit of West Spanish Peak (13,626 feet),* total vertical relief exceeds 7600 feet. Its northern boundary is Huerfano Park at the base of the Wet Mountains. Southeast of the Spanish Peaks, mesas arise abruptly above plains to a height of 9627 feet, the summit of Fisher's Peak. Much of the basin was covered with thick flows of basalt lava between 2 and 8 Ma.

In greater detail we begin with igneous features that were intruded into this region at the Oligocene-Miocene boundary, about 22.5 Ma. These events occurred widely and at the same time that a flare-up of caldera-forming eruptions in Nevada's Basin and Range province produced voluminous welded tuffs. These events are probably related to mantle instability at the completion of subduction events at the western continental margin.

IGNEOUS ROCKS

Intrusive Bodies

STOCKS

The Spanish Peaks. About 23 million years ago, the Spanish Peaks began as subterranean magma intrusions, also called *stocks* or *plutons*, that did not reach

*This approximation could be lower. See p. 7.

the surface before they cooled. Therefore, they were never volcanoes and did not produce lava flows or ash beds. The surrounding Tertiary sediments were domed up and altered (baked) in a zone hundreds of feet wide around the intrusion of the stocks.

The stocks cooled and crystallized as granitic masses. Later, after cooling, more uplift may have occurred after the Rockies arose from faulted anticlines. Late Tertiary erosion of the surrounding softer sedimentary rock caused the hard and solidified stocks to stand 1⅓ mile above the Great Plains to the east. Today, the boundary between Huerfano and Las Animas counties runs south of the summits and the saddle between them.

East Spanish Peak is mainly granite porphyry, with a large northwestern extension that terminates in a sill. Dated at 21.7 ± 1.0 Ma, the stock intruded strata of Tertiary age, causing them to dome up on the west and south. Its center is granodiorite porphyry, intruded slightly after the granite, which was chilled along its southeastern side against granite porphyry (Fig. 14).

West Spanish Peak is syenodiorite porphyry, dated at 22.9 ± 2.0 Ma, which intruded sedimentary rocks of Tertiary age and altered them by contact metamorphism. On West Peak's west-facing exposure, conspicuous horizontal lines represent remnants of contact metamorphosed sediments adhering to the

Fig. 14: View from near the summit of West Spanish Peak, with East Spanish Peak in view to the northeast.

intrusion. This zone is about 900 feet wide surrounding the intrusion (Fig. 15). The eastern and main mass of the stock crosscuts adjacent sedimentary rocks. On the west end, the stock intrudes its host rocks as a sill (laccolith). It caps metamorphosed sedimentary rocks and basaltic sills near the top of the peak. The base cuts across these rocks at a very low angle.

Rocks that were domed by the intrusion of the East Spanish Peak magma were later invaded by West Spanish Peak magma that caused no further structural deformation. It is believed that these stocks are partially supported, even raised a little, by dikes that were intruded shortly afterward.

Silver Mountain (Dike Mountain of Earlier Maps). Silver Mountain is an irregularly shaped stock of syenodiorite porphyry that was emplaced about 36.5 Ma. The surrounding contact rocks of Early Tertiary age are mostly covered.

Rough Mountain. Just north of Mount Maestas, Rough Mountain is an elongate stock composed mainly of syenite porphyry but also syenodiorite porphyry in places. Talus slopes and landslide debris cover the margins of the stock except at the southernmost exposures, where the base is nearly flat. Adjacent shale was metamorphosed to a distance of nearly 750 feet. Several small bodies of syenite porphyry that may be offshoots are a short distance from Rough Mountain.

Fig. 15: Climbing the south ridge of West Spanish Peak at about 12,000-foot elevation. Note the baked horizontal sediments that adhere to the surface. Numerous dikes radiate from the peak, and the Blanca Massif is seen in the distance to the west.

West of Rough Mountain is a small stock of syenite porphyry that cuts across sedimentary rocks and also across a thrust fault. The mass is subparallel to beds of Pennsylvanian and Permian ages in the upper thrust plate ("hanging wall") but cuts across the bedding of the Upper Cretaceous rocks below the sole ("footwall") (Fig. 16).

DIKES

The Spanish Peaks are world renowned for the presence of massive rock walls that are conspicuous from virtually all trails and roads in the region (Fig. 17). These are dikes, as many as 400 according to one count, formed as molten magma intruded by cutting across the bedding planes of the host rock. They are wall-like bodies of igneous material, mostly 10 to 50 feet thick (up to 100 feet or as little as 2 feet), that sometimes follow an irregular course to a maximum length of 14 miles. From the air, many appear to radiate like spokes from the Spanish Peaks (Fig. 18). They intruded as molten rock about 19.8 ± 1.6 Ma (or 26.6 to 21.3 Ma), vertically or at some other angle, into fractures in older layers of the surrounding sedimentary rock. Most are composed of syenite or monzonite (syenodiorite) porphyry, attached at their margins to metamor-

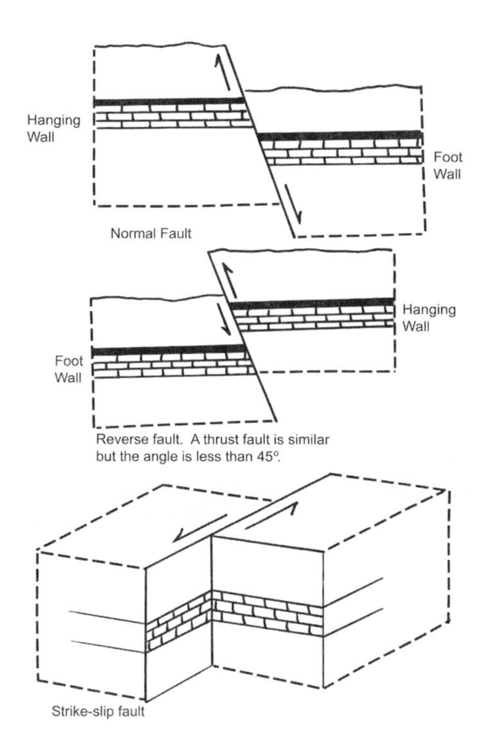

Fig. 16: Fault diagrams. A fault is a crack or break in the earth's crust. Movement along a fault produces an earthquake. The Sangre de Cristo normal fault, slipping a few feet at a time, has displaced the Sangre de Cristo Mountains more than 20,000 feet over the San Luis Valley. The range is what remains as ongoing erosion fails to keep pace with fault movement.

Fig. 17 (above): Dike emerging from the forest (right center) at about 11,000 feet, on the south-approach trail to West Spanish Peak. The southern Sangre de Cristo Range frames the distance.

Fig. 18 (left): Wahatoya Canyon. The dike in the foreground radiates from East Spanish Peak, seen on the right horizon.

phosed adjacent sediments. Most sills and dikes in other parts of the region were intruded at about the same time.

Dikes are generally more resistant to erosion than the intruded formations and consequently stand as nearly straight vertical walls as high as 100 feet. Some of the largest, radiating from West Peak, are easily seen on the east side of Route 12 south of La Veta, where the walls reach as high as 100 feet. The Devil's Staircase is a radial dike made of granodiorite porphyry along this route.

Mafic Dikes. A few softer dikes also exist that weather and erode more rapidly than the adjacent baked sedimentary rocks and form trenches. Called mafic dikes, they are high in iron and magnesium and low in silicon (similar to the composition of basalt), subject to erosion, and only infrequently found. Northeast of the Devil's Staircase is a pair of resistant dikes intruding the Cuchara Formation. Between them (now eroded away) is a lamprophyre dike intruded at 1000°C to 1100°C, baking the surrounding sandstone. Continuing on Route 12 south of Cuchara, a road cut on the west side through Permian red beds exhibits a vertical gray band that forms a shallow and narrow trench. This mafic dike is bounded by harder sandstones and conglomerates of the Sangre de Cristo Formation (Fig. 19).

Dikes of the Raton Basin region can be divided among three groupings, or *swarms*, as geologists may call them. Swarm one is composed of dikes radiating from the Spanish Peaks, as described above. The greatest concentration is seen west of West Spanish Peak, and they are generally short. They are conspicuous and tall along Route 12 between La Veta and Cuchara. On the eastern side of the peaks, the dikes are longer and lower. This swarm of dikes is mostly composed of syenodiorite porphyry with others of gabbro, diorite, syenite, granodiorite, and granite.

Swarm two is a series of subparallel dikes not associated with plutons. They are oriented northeast to southwest, a trend that follows the general strike of the folded sedimentary rock. They are as long as 14 miles, are the longest in this region, and are variously composed of syenite, porphyry, syenite lamprophyre, olivine gabbro lamprophyre, or feldspathoidal gabbro lamprophyre.

Swarm three is composed of dikes radiating from Silver Mountain (also called Dike Mountain locally). As with the Spanish Peaks, the dikes on the west are shorter than those on the east. They are similar in composition to the rocks that make up Silver Mountain and the Black Hills. They include textural varieties of syenodiorite. In addition, throughout the region there are many dikes, as well as small sets of dikes, that appear unrelated to other known intrusive bodies or dike systems.

SILLS

These originally molten intrusions are not basically different from dikes except that they were intruded along the planes of bedding within sedimentary strata. A good example of a low sill can be seen along a cut on the west side of the Indian Creek Trail, about a half hour south from the northern trailhead. It is composed of gray rocks intruded between red sandstone layers.

South and west of the Sheep Mountain and Little Sheep Mountain are thick sills of syenite porphyry and nepheline-syenite porphyry. They intrude overturned sedimentary rocks that range in age from Pennsylvanian to Late Cretaceous.

In various areas of the Raton Mesa region, sills are found within coal-bearing formations and commonly form ridges and ledges. Often they are composed of anastomosing stringers of igneous rock in narrow zones rather than as solid, thick sheets.

Fig. 19: In a road cut along the west side of Route 12, just south of Cuchara, is a mafic dike radiating from West Spanish Peak. Parking is possible on the east side of the road. The dike itself erodes faster than the adjacent Sangre de Cristo Formation. Note the adjacent, resistant country rock that was baked during the dike's intrusion.

White Peaks. North of Cuchara Pass, a north-to-south–trending ridge, 3 miles long and 2200 feet thick, is situated between West Spanish Peak and the Dakota Wall. It consists of the three White Peaks, North, Middle, and South, together constituting a sill of granite porphyry (also called "the Three Sisters" locally) (Fig. 20). Intruded during the Miocene (about the same time as the Spanish Peaks were emplaced), the sill entered sedimentary rocks of Late Cretaceous and Early Tertiary ages. At the north end the ridge is split into two sills. Immediately adjacent sedimentary rocks are apparently not deformed but are slightly altered near the intrusion.

LACCOLITHS

Similar to sills, these are flat-bottomed intrusions inserted within strata along bedding planes. Their domed tops cause an upward bulge in overlying sediments.

Little Black Hills. This small cluster of hills (also called the Black Hills), rising above the plain on the east side of 520 Road, were emplaced about 36.5 Ma. The hills are formed of syenodiorite porphyry that was intruded between shale beds of Late Cretaceous age and conglomerate and sandstone beds of Early Tertiary age. The overlying rocks were arched and baked. Nearby, several small isolated outcrops are probably offshoots of the main mass. Their base is generally well exposed, and the underlying rocks have been metamorphosed.

Sheep Mountain and Little Sheep Mountain. These peaks are formed of one large laccolith of microsyenite and syenite porphyry that intruded sedimentary rocks of Late Cretaceous and Tertiary ages. A small outlying mass at the south end has been separated from the main body by erosion. The base of the laccolith has a nearly planar surface inclined a few degrees to the northwest. Sedimentary rocks below the base are covered with talus and landslide debris, and the degree of metamorphism and deformation cannot be seen.

West of Sheep Mountain are three small stocks of nepheline-syenite porphyry intruded into overturned sedimentary rocks. Two of them intrude Jurassic rocks nearly parallel to the bedding plane. The westernmost stock intrudes or follows Pennsylvanian and Permian beds.

SOLE INJECTIONS

Mount Maestas. This mostly felsic stock is an elliptical mass of igneous rock that was injected along the plane of a thrust fault between shale beds of Late Cretaceous age, and overlying conglomerate beds of Pennsylvanian and Permian ages (Fig. 21). The mass is composed mainly of microsyenite and, locally, microgranite. The sole (footwall) of the thrust has been exposed by erosion,

Fig. 20: West Spanish Peak from the Dodgeton Trail. The White Peaks are seen in the middle ground.

Fig. 21: Timberline slopes of West Spanish Peak, looking northwest toward Mount Maestas.

and the underlying shale has been altered to phyllite. Southeast of Mount Maestas a linear body of basalt has been injected along the plane of a thrust fault. It does not appear to be connected to the larger body of Maestas itself.

A Bureau of Land Management (BLM) earth mover employee repeated a commonly heard story that 10 feet into the rock base, beneath fractured granite, is blue ice. Hard ice is not uncommon beneath rock glaciers, the somewhat unstable surface rocks covering the slopes of stocks in this region.* Although Maestas is a relatively low peak for this region, it is considered a more dangerous climb than some fourteeners. David Steffan, a local historian, has said that it is like climbing a stack of dinner plates.

PLUGS

Goemmer's Butte. Named for the Goemmer family that homesteaded the area in the 1860s, this conspicuous volcanic feature is a latite (trachyandesite) plug that intruded the Cuchara Formation, an Eocene arkosic conglomerate. The feature preserves the conduit or throat of a small volcano (e.g., a cinder cone), whose surface deposits have been eroded away. Several blocks of sandstone are contained within the body of the plug, a testament to the power of the rising magma to remove large blocks of country rock and carry them upward. Irregularly elliptical, the plug is about 900 feet across in its widest dimension (Fig. 22). The surrounding rock was not extensively altered, but remnants of bleached sandstone adhere to the wall of the plug. Latite is a dark porphyritic rock, intermediate between trachyte and andesite (extrusive equivalent of monzonite). Little or no quartz is found within the crystalline ground mass. There are scattered medium- to coarse-grained mafic intrusions.

Goemmer's Butte is unique in the area as the only place where magma vented to the surface, from a central fissure or conduit. It likely postdates the intrusions of East and West Peak stocks.

Huerfano Butte. The county's eponymous feature is just to the east of I-25, about 9 miles north of Walsenburg. It is a conspicuous little knob, rising about 300 feet above the surrounding plain (Fig. 23). The butte is a plug composed of biotite olivine alkali-gabbro and bisected by two east-trending dikes, one of biotite monzonite (syenodiorite) and the smaller of alkali lamprophyre. The

*Whether the ice is kept frozen by the proximity of naturally occurring CO_2 in this area is uncertain. A large reservoir of naturally occurring CO_2 beneath the neighboring Sheep Mountains has been extracted for years by energy companies. The CO_2 is piped to the Permian Basin oil fields of Texas, where it is used to enhance oil recovery. When ancient buried vegetation decays, CO_2 is the first gas to accumulate from the breakdown process.

Fig. 22 (above): Goemmer's Butte, looking south from Sulphur Springs Road.

Fig. 23 (left): Huerfano Butte, "the Orphan," north of Walsenburg, just to the east of I-25. The county was named after this igneous feature.

feature originated about 26 to 25.2 Ma (Oligocene), making it earlier than the emplacement of the alkali intrusive rocks of the Spanish Peaks. Among other travelers, John C. Fremont's last western expedition, a railroad survey, noted the butte in 1853 on its way to California.

Smaller Examples. Four miles southeast of Little Black Hills is a microgranite plug that intruded sedimentary rocks of Tertiary age. It is elliptical and less than 100 feet across. The mass is truncated by a pediment and is not well exposed.

Northwest of Walsenburg is a small plug of scoriaceous basalt with inclusions of gneissic cobbles that cut shale beds of Late Cretaceous age. Its perimeter is full of gas bubbles (vesicles), indicating that the preserved rock was emplaced and cooled near the earth's surface. Isolated plugs or groups of plugs east and northeast of Walsenburg intrude shale beds of Late Cretaceous age. They are elliptical in outline and as much as 450 feet across. The rocks vary from porphyritic diorite to syenodiorite porphyry. Heat from the magma caused a small amount of alteration of the enclosing sedimentary rock.

Extrusive Bodies

LAVA FLOWS

The Raton Mesa's Colorado/New Mexico border region was capped by lava flows originating in northern New Mexico, probably within the Raton-Clayton volcanic field.* Within Colorado, erosion has removed most of these hard caps and much of the softer underlying sediments, but remnants remain in a few places. A most conspicuous example is Fisher's Peak, just east of Trinidad, with its flat lava cap resulting from a magma flow that ended about 3.5 Ma (Fig. 44). Beneath the summit an additional seven sheets of lava can be distinguished. As many as 11 separate flows (occurrences) make up the total lava cap of the Raton and Barilla mesas. Each flow may be up to 30 feet thick and is composed of dark-gray olivine and locally vesicular basalt. Commonly, each flow is scoriaceous near its upper margin. The total lava thickness for Raton Mesa may have been 500 feet.

VOLCANIC CONES

In the southernmost part of the Wet Mountains (the northern edge of the Raton Basin) is a cinder cone. Also, just west of Rough Mountain, a small spat-

*Capulin Volcano National Monument in northeastern New Mexico is located near the center of this field.

ter cone has erupted onto Late Cretaceous sedimentary rocks. It is less than 50 feet in diameter with rock composed of scoriaceous basalt.

NON-IGNEOUS FEATURES

Sangre de Cristos

This range is one of the longest on Earth, extending from Poncha Pass, Colorado, to Glorieta Pass, New Mexico. A formidable though narrow rampart throughout its length, it contains 10 peaks higher than 14,000 feet and 25 peaks higher than 13,000 feet. A natural divide separates the northern portion of the range from the southern, which is also called the Culebra Range. At La Veta Pass, Highway 160 follows this natural divide. The northern range's terminus rises dramatically from the plain just north of Fort Garland as the great Blanca Massif that contains four fourteeners (Fig. 24). This great collection of peaks that exposes Precambrian rocks is bounded on the east, west, and south by high-angle normal faults. Block rotation has offset the northern half of the range with respect to the Culebra Range by about 25 miles, roughly along Route 160 from La Veta Pass to Fort Garland.

Fig. 24: The Huerfano River emerges from the upper Huerfano Basin and drains the north side of the J-shaped range of the Blanca Massif.

The main body of the Sangres Range is of Permian-Pennsylvanian age, a mix of igneous intrusions, conglomerates, and shales originating about 250 Ma (Fig. 25). Not just a large anticline, this range was originally elevated by low-angle thrust faults during the Laramide deformation when the Rockies were laterally compressed in Late Mesozoic to Early Tertiary time. By Mid-Tertiary (beginning 27 Ma), the bordering fault systems of the Sangres were reactivated and the range was uplifted as a block

Fig. 25: Looking south from Bear Lake Divide toward Trinchera Peak. This peak, with its long-persisting, north-facing snow field, marks the southwest corner of Huerfano County.

(horst). Valleys (grabens) dropped on the west (San Luis Valley) and east (Wet Mountain Valley) sides.

The Sangre de Cristo normal fault facing the San Luis Valley is currently active. It is estimated that movement during the last 50,000 years has averaged 0.17 mm/year with an average recurrence of slip every 12 years. More than once in the last 15,000 years its movements have produced earthquakes of magnitude 7.0 to 7.5.

TEDDY'S PEAK (~11,960 FEET)

As described here, Teddy's Peak will serve as an example of the geology of the Culebra Range that looms over the former Cuchara Mountain Resort. From the trailhead above Bear Lake (see Chapter 8, Trails, Teddy's Peak) to the top of the ridge is arkosic sandstone and conglomeric sandstone of the Permian/Pennsylvanian Sangre de Cristo Formation as mapped by Lindsey (1995). These rocks are generally covered by soil and loose rock that mantle the base of the range. In the saddle south of Teddy's Peak, the rock changes to coarse granite, dominated by orthoclase, and gneiss. The rocks contain quartz stringers and coarsely crystalline (pegmatitic) zones. This granite contacts dark-gray, strongly foliated (layered) hornblende gneiss on the north end of the saddle

(~11,920 feet). Foliation trends along the ridge (roughly north-south), with a steep plunge (~70-degree angle east).

The granite appears to postdate the gneiss, as indicated by granitic stringers within the gneiss, often intruded along foliation. Further up the ridge, to the north, beneath the southern minor peak of Teddy's Peak, one encounters granitic gneiss intruded by granite veins with well-developed foliation. This also forms the summit of Teddy's Peak. The ridge south of the approach appears to be composed of near vertically foliated dark-gray gneiss.

The western flank of the range as seen from the peaks is formed of the Mid-Permian Madera Formation, overthrust by basement crystalline rocks of Early Proterozoic age, about 1.6 Ga. They are formed of felsic gneiss, augen gneiss, younger felsic pegmatites, and mafic dikes. The Permian rocks comprise 3000 feet of gray limestones interbedded with red arkosic sandstone, siltstone, and shale.

TRINCHERA PEAK

This is the northernmost high point of the Culebra Range (Fig. 25). It is formed of the sandstone bedrock of the Sangre de Cristo Formation, vertical to slightly overturned. Its basal sandstone is erosion-resistant arkosic material that causes the peak's prominence. Derived from alluvial fan sediments laid down in Pennsylvanian-Permian times, it accumulated as erosion products of the Ancestral Rockies. These originally horizontal strata were elevated in the Mid- to Late Miocene by processes of uplift, faulting, and block rotation.

When driving west on Route 12 up the Purgatoire Valley, the southern Sangres (Culebra Range) appear as a jagged line of peaks. Faulting tilted and raised Paleozoic sediments that extend to the top of the range. Pennsylvanian rocks form many of the summits. About 20,000 feet of Permian sandstone forms the upper part of the mountain slopes.

The Dakota Wall

This formation of hard sedimentary quartz-rich sandstone was deposited about 100 Ma as a Cretaceous beach. It is the oldest unit of that period in this region (Figs. 26 and 39). Its sediments accumulated under a marine transgression that stretched from the Gulf of Mexico to the Arctic. After lying under other sediments for perhaps 40 million years, the Dakota sandstone was elevated during the Laramide orogeny (65 Ma). At that time, this very hard bed was broken and pushed up at an angle. After much later erosion, sections of the Dakota sandstone remained. Running north-south, the formation forms a recognizable landmark at various places between Mexico and the Canadian border.

Fig. 26: The Dakota Wall crosses Route 12 near Stonewall and courses southward.

Just to the west of the Spanish Peaks the Dakota Wall is a resistant, nearly vertical hogback that dips steeply eastward. The softer adjacent layers to the west (the Cuchara Valley) were tilted at the same angle and eroded to a lower level. It demonstrates several hundred feet of reactivated Laramide fault movement that originated in Late Paleozoic. The wall is conspicuous along Route 12 from Stonewall in the Purgatoire Valley to the south, to 15 miles north of Cuchara Pass. The upper Cuchara Valley ends on the north end, where the wall crosses from the east to the west side of Route 12 at a point called "the gap." A trail, inaccurately called the "Dike Trail," follows the east side of Cuchara Valley, roughly following the Dakota Wall and providing access to it (see Chapter 8, Trails).

Farther afield, sections of the Dakota Wall are visible aboveground on the east side of the Wet Mountain Valley and western foothills of the Wet Mountains. Other notable exposures of the Dakota Wall are at Garden of the Gods at Colorado Springs and to the west of Boulder, Colorado.

The Cuchara Pass Region

This pass (elevation 9941 feet) is a saddle between the Spanish Peaks and Sangre de Cristo Range and forms the western boundary of the Raton Basin. The road east to Cordova Pass exposes a sequence of rocks from Upper Paleozoic to Tertiary strata. (These strata can also be seen at I-25; see below.) Driving east from Cuchara Pass toward Cordova Pass (gaining elevation) one travels upsection* over beds that dip steeply to the east. These are:

1. Upper Cretaceous limestones, shales, and chalk of about 3500 feet total thickness. The top 2200 feet is the Pierre Shale.

2. Trinidad sandstone and Vermejo Formation, upper portion of the Upper Cretaceous. The rocks are arkosic sandstone, siltstone, shale, and coal to a total thickness of 487 feet.

3. Poison Canyon and Raton formations. These uppermost Cretaceous and Paleocene sediments are arkosic sandstone, siltstone, and conglomerate to a thickness of 2000 to 2200 feet.

4. Cuchara Formation. This formation, of Eocene age, is composed of poorly consolidated arkosic sandstone and conglomerate, locally metamorphosed to slate and quartzite from contact with Spanish Peaks intrusions. In the vicinity of Cordova Pass, two members are distinguished: (1) conglomerate interbedded with sandstone and mudstone (about 2600 feet thickness) with rounded cobbles and boulders of Proterozoic and Paleozoic ages from the Culebra Range; and (2) a sandstone member, interbedded with mudstone and poorly consolidated conglomerate to a total thickness of about 1000 to 1800 feet. Locally it contains logs, stumps, and wood fragments.

The Cuchara Formation was accumulated as erosional fans and river deposits from the Culebra Range after the Laramide orogeny. The west side of the Raton Basin was level at the time, as attested by the horizontal banded appearance of the Spanish Peak summits. The banding is baked metamorphosed fragments of the Cuchara Formation.

Looking west from Cuchara Pass, the entire sequence from pass and foothills to the peaks is the Sangre de Cristo Formation, whose rocks are often blood red (Fig. 27). Its age is Pennsylvanian to Lower Permian, and it constitutes a record of sediments eroded from the Ancestral Rockies. Red beds are a common indicator of Permian rocks, which can be found in the southern Plains (the Permian Basin) in New Mexico, Texas, Oklahoma, and western Kansas. The Sangre de Cristo beds are composed of arkosic sandstone, siltstone, shale, and conglomerate, with crossbedding, ripple marks, channels, and cracks. Conglomerates indicate deposits from rapid erosion close to the source. Just south of Cuchara on Route 12, road cuts expose these red beds, which dip steeply eastward across the Cuchara Valley. Higher in the foothills are numerous folds, normal faults, and thrust faults.

*This refers to the direction of travel over increasingly younger strata.

Fig. 27: The Dodgeton Creek Trail runs across the red rock of the Sangre de Cristo Formation.

Fig. 28: Crystalline rock on the slopes of Greenhorn Peak on the trail that runs from the parking lot to the east. View looking west toward the upper Huerfano Valley. Blanca Massif in the right distance.

Wet Mountains

Straight north of La Veta and dominating the skyline west of I-25 between Pueblo and Walsenburg is Greenhorn Peak (elevation 12,334 feet), the summit of the Wet Mountains (Fig. 28). The range is a west-northwest–trending block that merges with the Front Range near Royal Gorge. It is defined by the Wet Mountain thrust fault on the northeast side and by normal faults on the west and south sides. Late Cenozoic normal faults define the southeast side.

Much of the range is formed of Proterozoic crystalline rock that was uplifted along flanking faults on the east and west. The rocks are migmatites, amphibolites, gneisses, gneissic-granodiorite, and granites. A remnant of Oligocene volcanic rock caps the Precambrian unconformably near the summit. This is the Rocky Mountain (Eocene) erosional surface of 27 to 38 Ma. On Greenhorn Peak, this surface preserves a basaltic andesite flow of 33.5 Ma.

Late Tertiary displacement along the Ilse Fault, which trends northwest to southeast within the range, shows that the modern Wet Mountain block was most recently uplifted within the last 24 million years. To the west, the Wet Mountain Valley is a graben. Faults along its east and west sides allowed valleys to descend as mountains arose.

AT THE EDGE OF THE GREAT PLAINS

STRATIGRAPHY

Heading south from Walsenburg on I-25, the 37-mile drive to Trinidad displays a panorama characteristic of the Raton Basin at the western margin of the Great Plains. These strata, which underlie much of the Great Plains, are warped upward and exposed at the mountains. They also define the eastern limits of the region's coal fields.

The highway itself lies on the Pierre Shale. This Upper Cretaceous shale and siltstone deposit, which is as thick as 300 feet, grades into fine sandstone and includes limestone concretions. Looking west and sequentially upward, we see an escarpment or small cliff (the pinyon-juniper zone) of the Upper Cretaceous Trinidad Formation, a fine-grained sandstone. It is marked by casts of burrows formed by *Ophiomorpha*, a shrimp-like organism. With changes in sea level the shoreline fluctuated, causing some interfingering of beds.

Next upward, we find another sandstone, the Upper Cretaceous coal-bearing Vermejo Formation. It is composed of fine- to medium-grained sandstone interbedded with mudstone, shale, and thick coal beds, top and bottom. Next is the Raton Formation, which extends from Upper Cretaceous through Lower Paleocene. It contains a lower coal zone, overlain by fine sandstones and mudstone, and an upper coal zone. Just above the lower coal zone, evidence is found of the abrupt Cretaceous-Tertiary (K-T) boundary that defined the end of the Mesozoic era.

Above this is the Upper Paleocene Poison Canyon Formation, formed of up to 50 feet or more of sandstones and conglomerates. To the west it thickens and inter-tongues with the Raton Formation.

FISHER'S PEAK

Looking south as one approaches Trinidad, Fisher's Peak looms over the town from the east. This formidable mass is distinguished by its flat summit formed of erosion-resistant basalt from an eruption that ended about 3.5 Ma. About 800 feet of basalt overlies the Miocene Ogallala Formation, which, in turn, rests on the Cretaceous Dakota sandstone. Both are permeable to water. The

Ogallala is famous as an aquifer that underlies 174,000 square miles in eight Great Plains states. It is said to store 3 billion acre-feet of water upon which regional agriculture depends.

Fisher's Peak basalts are divided among 11 different flows that originated from a volcanic vent to the south. Some geologists believe these flows are remnants of a very widespread flow when the entire region was at that elevation. These deposits constitute the most direct evidence of extensive Cenozoic erosion that lowered most of the surrounding landscape by about 3700 feet in the past several million years.*

A fragment of this basalt, a hogback, can be seen just east of I-25, south of the Ludlow exit. This dark feature standing abruptly above the plains has been fancied to resemble the silhouette of a 1920s Franklin automobile, at least the outline of its hood and trunk.** Whether it represents the same flow as that on Fisher's Peak and Bartlett Mesa has not been confirmed.

The American West is noteworthy for its accumulation of dramatic geologic features, all the more because the semi-arid climate exposes so much of the earth for our viewing convenience. The Spanish Peaks and their surroundings, a relatively compact area, are second to none for geologic variety and would be a worthy attraction in the absence of any other reason.

*This is a modest rate of erosion. At a loss of 1 inch per century, more than 4000 feet of sediments would be removed in 5 million years. Much of our landscape today is eroding at a far greater rate.

**Another view states that this structure is actually a dike.

GEOLOGIC GLOSSARY

Alkali—A descriptive term for volcanic rocks rich in sodium and potassium.

Alluvial fan—A fan-shaped pattern of sediments deposited by steam flow or floods, usually at the mouth of a valley or edge of a mountain range.

Amphibole—A complex group of silicate minerals, igneous and metamorphic (e.g., hornblende) with Ca, Fe, Mg, SiO_2, and OH_2; a metamorphic rock formed predominantly of amphibole minerals.

Anastomosing—A braided pattern; for example, intruded veins of rock, or streambeds that separate and rejoin.

Andesite—Gray to black volcanic rocks with intermediate (52% to 63%) silica (SiO_2) content, crystals of plagioclase feldspar, and one or more of the following: pyroxene, hornblende, and maybe olivine (when at the lower end of silicon content). Andesite magma usually forms thick lava flows, often from large and explosive eruptions of ca. 900°C to 1100°C; multiple outpourings of andesite form strato-volcanoes.

Anticline—Folded rock formation, upwardly convex.

Arkose (arkosic)—Sandstone of continental origin, mostly of quartz and orthoclase feldspar, usually coarse grained and pink.

Augen—Lens- or eye-shaped features in metamorphic rocks (such as "augen gneiss") formed by coarse-grained crystals surrounded by fine-grained crystals that appear to flow around the larger ones. Inclusions are often of feldspar, garnet, or quartz.

Basalt—Rock formed from the cooling of mafic lava, which is low in silica and relatively high in Fe and Mg; common minerals include olivine, calcium-rich plagioclase, and pyroxene. This lava flows at ca. 1100°C to 1200°C, with low viscosity and often forms miles-long, thin lava flows, as one can see erupting in Hawai'i.

Biotite—Component of silicic rocks (granite, rhyolite): K, Mg, Fe+++, Al, SiO_3, OH, F. = iron mica, monoclinic, prismatic.

Block rotation—Faulted blocks that slip laterally with respect to one another; a typical result of transform faults.

Canadian shield (see Craton)

Cirque—A bowl-shaped head of a valley, usually in the alpine zone, carved and shaped by a glacier.

Conglomerate—Sedimentary rock composed of larger fragments (gravel and cobbles) embedded in a finer matrix (e.g., sand, silt). Characteristic of rapidly accumulating sediments deposited close to an erosional source.

Craton—The stable core of a continental plate, usually composed of ancient plutonic and metamorphic rock. For example, the Canadian Shield of North America.

Crust—The outermost layer of the earth, 5 to 25 miles thick. Oceanic crust is composed mostly of basaltic igneous rock, some metamorphic, and a thinner layer of sediments. Continental crust is composed mostly of silicic rock (e.g., granitic plutons), which is less dense, and a lower zone of denser mafic rock.

Crystal size—Intrusive rocks have large crystals (coarse grained) due to slow cooling. Extrusive rocks have small crystals (fine grained) formed by rapid cooling.

Crystallization—Formation of mineral bodies that have facets and straight edges due to an ordered, internal atomic structure.

Diabase—Dark-gray mafic rock intruded at shallow depths, cooled at a rate intermediate between rapid (e.g., basalt) and slow (e.g., gabbro).

Diorite—Intrusive igneous rock, intermediate between mafic and felsic; this rock is often recognizable by its black-and-white, salt-and-pepper appearance.

Earthquake zone—Earthquakes release stresses along faults that deform portions of the shallow crust. Such stress can be due to compression or extension. The quakes usually occur in the outer brittle portions of the crust, where it is thinner and rocks are relatively cool. Strain stores tremendous energy that is released suddenly.

Escarpment—The side of a cliff formed by erosion or faulting, usually separating two relatively level areas.

Extrusive—Molten rock that is erupted and solidifies on the earth's surface.

Fault—Breaks or cracks in the crust where blocks move with respect to one another in response to stress. Three types are common: ***normal***, vertical movement; ***thrust***, vertical, where older beds override younger beds; and ***transform*** or ***strike-slip***, where strata move laterally with respect to each other. (See Fig. 16.)

Feldspar—Felsic mineral whose name comes from "field spar." As many as 20 types of this easily cleaved mineral exist. It is common in intrusive igneous rock with quartz and mica and often with silica and aluminum (e.g., granite). *Plagioclase* (six types): Andesine, Na, Ca, Al, SiO; with triclinic, oblique cleavage planes, not at right angles; some with striations. *K-feldspars*: K, Al, SiO. or K, Na, Al, SiO; low symmetry, monoclinic or triclinic, twin easily. Crystallizes in igneous environment; also found in metamorphic rocks. Has two easily cleaved planes. *Orthoclase*: Right-angle cleavage. Pink rock, if light, with no striations.

Fell fields—Rocky alpine ground caused by freezing and thawing cycles that, over time, carry rocks to the surface, forming distinct patches of cobbles and boulders.

Foliation—Layering in metamorphic rock imparted by mineral banding.

Gabbro—Mafic igneous intrusive rock, usually dark colored. Comparable in chemical composition to basalt (volcanic). Slowly cooled, coarse grained; commonly contains calcium feldspar, pyroxene, olivine, and rarely quartz.

Gneiss—Metamorphic coarse-grained foliated rock. Minerals form in discrete layers under pressure and chemical activity.

Graben—A down-dropped block, due to faulting, adjacent to a raised block (see Horst).

Granite—Coarse-grained igneous plutonic rock, light colored, typically with quartz, orthoclase feldspar, and mica.

Granodiorite—Igneous, plutonic rock, less felsic than granite, light colored. Comparable to the extrusive rhyodacite, which is light-gray volcanic rock with plagioclase, K-feldspar, or other crystalline minerals in glassy silica.

Hogback—Ridge with steeply sloping sides that may be formed of steeply tilted resistant strata.

Hornblende—The most common form of amphibole minerals (see Amphibole).

Horst—A raised block, due to faulting on flanking normal faults, adjacent to a dropped block (see Fig. 16) (see Graben).

Igneous—Rock solidified from magma.

Intrusive—Molten rock injected into existing rock and cooled below the earth's surface.

Laccolith—Similar to a sill, a flat-bottomed intrusion between bedding planes in sedimentary rock. Its domed upper surface causes overlying sediments to bulge upward.

Lamprophyre—Fine-grained, dark-colored igneous intrusive rock with feldspar and a high proportion of ferro-magnesian (iron- and magnesium-bearing) minerals.

Latite (*syn.* trachyandesite)—Volcanic rock characterized by phenocrysts (crystals) of plagioclase and K-feldspar in approximately equal amounts. Volcanic equivalent of monzonite (*syn.* syenodiorite).

Mafic—Dark-colored igneous rocks rich in magnesium and iron (*MA*gnesium + *Fe* = "mafic") and low in silica.

Magma—Molten rock, either intrusive (plutonic) or extrusive (lava). *Mafic* magma has little free silica, is dark colored, contains high content of Mg and Fe and plagioclase feldspar, flows easily at ca. 1000°C to 1200°C, and cools as, for example, basalt (extrusive) or gabbro (intrusive). *Silicic* magma is up to 75% silicon dioxide (silica), crystallizes quartz and orthoclase feldspar ("K-spar"), and melts at about 600°C to 800°C, but is viscous and resistant to flow and therefore often erupts explosively. Cools slowly to form granite (intrusive) or rapidly to form rhyolite (extrusive). *Intermediate* magma is very common. It forms andesite lava, which melts at intermediate temperature (900°C to 1000°C) and flows freely. It may form as a result of mafic magma that reacts with silicic continental rocks as it ascends from the mantle. As plutons, it forms coarse-grained diorite and syenite.

Mantle—Earth layer beneath (internal to) the crust. At its upper boundary, earthquake waves abruptly change velocity due to a density contrast as the rock changes to more silicic crust. It is formed of ultramafic minerals, iron-magnesium silicates, and relatively low proportion of aluminum and feldspar.

Massif—A large mountain mass or cluster of mountains within a range (e.g., the Blanca Massif of four fourteeners).

Metamorphic—Rocks, originally igneous, sedimentary or other metamorphic, altered structurally and minerologically by pressure and heat.

Micro—As an adjective, meaning "small" or "very fine grained"; crystals not easily visible with the naked eye.

Migmatite—A rock composed of a mixture of metamorphic (e.g., schist and gneiss) and intruded igneous rock; often veined, streaked, or mixed in appearance.

Monzonite (*syn.* syenodiorite)—Igneous intrusive rock, with quartz, various feldspars, some biotite, and hornblende.

Moraine—Earth surface materials pushed or carried and left behind by a glacier, as on the surface (ground), along the side (lateral), or at the farthest extent (terminal) of the ice.

Nepheline—Igneous feldspathoid mineral composed of sodium potassium aluminum silicate; greasy luster; without cleavage; an essential constituent of some sodium-rich and silica-poor rocks (e.g., nepheline syenite).

Normal fault (see Faults)

Olivine—Silicate mineral (Mg, Fe, SiO_4), no cleavage planes, glassy. One of the first minerals to crystalize from molten mafic magma; a common constituent mineral of basalt and diorite.

Orogeny—Mountain building by some combination of folding, faulting, or regional uplift.

Orthoclase (see Feldspar)

Outwash terrace—Earth deposits left behind along the path of rapid runoff from a melting glacier. Usually distinguished from a moraine in having cobbles that are rounded or smoothed.

Pediment—A broad, sloping plain formed by erosion adjacent to the base of mountains or hills.

Pegmatite—Igneous coarse-grained rock, usually granitic; often a source of semiprecious gems.

Periglacial—Features formed near the perimeter of glaciers under the influence of low temperatures.

Phenocrysts—Large crystals formed in igneous rock.

Phyllite—Low-grade metamorphic rock intermediate between slate and schist, exhibiting lustrous sheen imparted by aligned, microscopic mica crystals; often formed from clay-rich sediments formed from mudstones and shales.

Plates—Fragments of the earth's crust that include the adjacent uppermost mantle (lithosphere). They are constituents of continents and larger islands. On the earth's surface are about a dozen large plates and a number of smaller ones with an average thickness of 50 miles.

Plate motion—Plates separate due to the welling-up of new crust from hot material in the deep mantle at mid-oceanic ridges. Hot and light at first, plates cool and become heavier as they move away from spreading zones. Plate separation on one side may be accompanied on the other by collision or subduction beneath adjacent plates.

Plate tectonics—Plates move relative to one another over a hotter, deeper, more plastic mantle at speeds of a few inches per year or about the rate that fingernails grow. There are three types of plate boundaries: *spreading*, where new material is added at a mid-oceanic ridge or continental rift zone (e.g., Rio Grande Rift, East African Rift); *convergent*, where a denser plate is subducted beneath a lighter one; and *transform boundaries*, where two plates slide horizontally past one another (e.g., the San Andreas Fault zone where the Pacific Plate slides northward relative to the western margin of the North American Plate).

Platform—The thinner crust closer to the margin of a tectonic plate.

Plutons—From Pluto, the Greek god of the underworld. Igneous rocks intruded as magma into existing beds and solidified beneath the surface. They are usually coarse grained as opposed to the finer-grained volcanic rocks. When mountain sized, they are also called *batholiths*; smaller plutons ($< 100\ km^2$) are called *stocks*.

Porphyry—Igneous intrusive rock, with two distinct crystal sizes: fine-grained groundmass contains large feldspar and quartz crystals.

Pumice (pumiceous)—Vesicular glassy rocks of rhyolitic composition.

Pyroxene—Silicate mineral group; dark crystals high in Mg and Fe; component of mafic and intermediate rocks.

Quartzite—Very common rock of metamorphosed quartz sand, which is mostly silicon dioxide.

Rift zone—A crack in the earth's crust that may widen over time. The earth's crust thins and drops down along normal faults at the margins while the mantle wells up along the axis.

The high heat flow associated with a rift causes volcanism in the surrounding region. A sign of crustal extension (e.g., the Rio Grande Rift).

Rocks, common igneous:

Volcanic	Plutonic	Silicon Content	Minerals in Order of Abundance
Basalt	Gabbro	Low	Pyroxene, olivine, feldspar, amphibole
Andesite	Diorite	Medium	Feldspar, amphibole, pyroxene, biotite mica
Rhyolite	Granite	High	Feldspar, quartz, biotite mica, amphibole

Scoria (scoriaceous)—Igneous, extrusive rock. Cooled mafic lava fragment (pyroclast) of basaltic composition, with abundant gas vesicles.

Sedimentary—Rock formed from accumulated particles eroded from an adjacent region of higher elevation. Sediment particles accumulate in marine, lake, or stream bottoms, or in terrestrial areas from stream or wind deposits.

Shale—Fine-grained sedimentary rock consolidated from silt and clay particles.

Sole injection—Igneous intrusion along a thrust plate.

Stock (see Pluton)

Stratum (strata)—Layer or bed of sedimentary rock.

Strike—The azimuth of a fault or tilted stratum where it intersects the earth's surface. Combined with the dip angle (perpendicular to the strike), these measurements provide a description of the orientation of a geologic feature.

Strike-slip, transform (see Faults)

Stringers—Small igneous veins that intruded irregular cracks of existing rock.

Subduction—An intersection of two tectonic plates in which one plate (usually denser) dives beneath the other. As the subducting (down-going) plate descends into the mantle, it heats up, releases fluids, and eventually melts. The liberated fluids cause local partial melting of the upper mantle, forming magma bodies that rise into the overlying crust.

Syenite—Igneous rock of alkaline feldspar and other minerals such as hornblende.

Syenodiorite (*syn.* monzonite)—Plutonic rock bearing various feldspars and ferromagnesian materials.

Tectonic activity—Breaking and bending of the crust under internal forces, usually as a result of plate collisions.

Thrust fault (see Faults)

Till, glacial till—Unstratified soil materials containing angular rock fragments, of many sizes, that were carried or pushed by glaciers.

Trachyte—Igneous rock containing abundant alkalis (Na + K), often with alkali feldspar.

Tuff; welded tuff (*syn.* ignimbrite)—Consolidated volcanic ash, pumice, and rock fragments originating as explosive volcanic eruptions and associated ash flows.

Unconformity, unconformably—Boundary beneath strata lying on an erosional surface, where the original sequence of deposition was interrupted.

Vescicles (vescicular)—Pockets formed by gases within molten rock (analogous to hollow pockets within rising bread). The structure of pumice and scoria.

Volcanoes—Edifices constructed of extrusive igneous rocks, such as lava flows and ash and pumice beds (pyroclastic deposits). In North America they develop behind plate boundaries above subducting plates. Typically they form above points where subduction reaches a mantle depth of 450 miles. Volcano locations demonstrate the angle of trajectory of a subducting plate. Volcanic magma originates about 80 miles deep below the edge of subduction.

CHAPTER 5

REGIONAL GEOGRAPHY AND ECOLOGY

GEOGRAPHY AND ECOLOGY

Trying to understand a landscape is a challenge, especially when encountering it for the first time. It yields best when focusing on one factor at a time. To begin, we can divide our territory into physical and biological components, even though the two are intertwined in so many ways.

ECOSYSTEMS

A large unit of land that includes its living species and the physical and competitive factors is called an *ecosystem*. Many are recognized in North America. A particular ecosystem is named for a dominant indicator species or habitat, such as the ponderosa pine ecosystem or the alpine ecosystem. It is both a unit of study and a living geographic space.

Life in an ecosystem is dependent first on the physical earth: its soils, water, solar radiation, weather, and climate. The communities of living organisms absorb energy from the sun and then pass energy from one species to another through food chains or webs. First, plants accumulate nutrients from the soil and energy through photosynthesis. Animals, the consumers, are dependent on plants. Fungi and microscopic bacteria are the decomposers that cycle huge amounts of nutrients through the system. The density of populations of all living things, and their diversity, in an ecosystem can be used as a direct measure of equability; that is, the generosity of conditions that support life.

The American West is an ideal location to be looking at ecology and geography because the number of players (individual species) in a given area are few enough that we can hope to understand the patterns. First, we will look at physical factors especially characteristic of the mountainous West, followed by biological factors, the ways in which living things respond and thrive.

FACTORS CONTROLLING PLANT LIFE

PHYSICAL FACTORS

1. *Elevation.* The region covered by this guide varies from about 5500 feet, east of Trinidad and Walsenburg, to just under 14,000 feet on West Spanish Peak. As one moves upward, growing conditions become more severe. Exposure to ultraviolet and visible light increases, as do the temperature extremes. Higher winds result in faster desiccation. Growing seasons are shorter, and soil development takes centuries.

2. *Radiation* is much more intense at high elevations. From sea level the sun's intensity increases about 4% per 1000 feet. At the western edge of the Great Plains there is 24% more solar radiation exposure, about 36% more at 9000 feet, and 56% more on top of West Spanish Peak. In addition to the harm that sun can do to our skin, too-great intensity also has the capacity to bleach (destroy) chlorophyll. High-elevation plants have evolved structures and pigments capable of deflecting excess radiation away from sensitive photosynthetic pigments.

A look at the figure on the following page will show that the woody plants of the region tend to assort into communities based on elevation. Just by looking at a photo that includes background vegetation, an experienced person can determine the elevation where the photographer was standing to an accuracy of less than 1000 feet. Most species have well-defined requirements for survival in nature despite the fact that, in cultivation, most species will survive well outside their natural ranges.

3. *Atmospherics: air and wind.* The atmosphere is considered a compressible "fluid" that consists of about 12 gases, mostly nitrogen, oxygen, argon, and CO_2. It may contain up to 4% water vapor in the most humid places on Earth but is very much less in arid areas. When gaining elevation, barometric pressure is reduced approximately 10% per half mile above sea level.

Wind, the atmosphere in motion, is rather constant in this region and its presence cannot be ignored. It is a direct result of the short- and long-wavelength radiation that reaches the earth from the sun. Some of the short-wave energy, or visible light, is converted to living matter through photosynthesis. Much is converted to long-wavelength energy, the heat that leads to wind. Solar heating of surfaces, such as the south-facing slopes of mountains, in turn heats the air column above them. When heated, the air becomes less dense, exerts pressure on air above, and rises. Cooler air near the surface moves in behind it. In turn, cool air drops from above, and a circulation cell develops.

The atmosphere can be considered a fluid and, as wind, it always moves toward uniform redistribution of energy and pressure. But the work is never

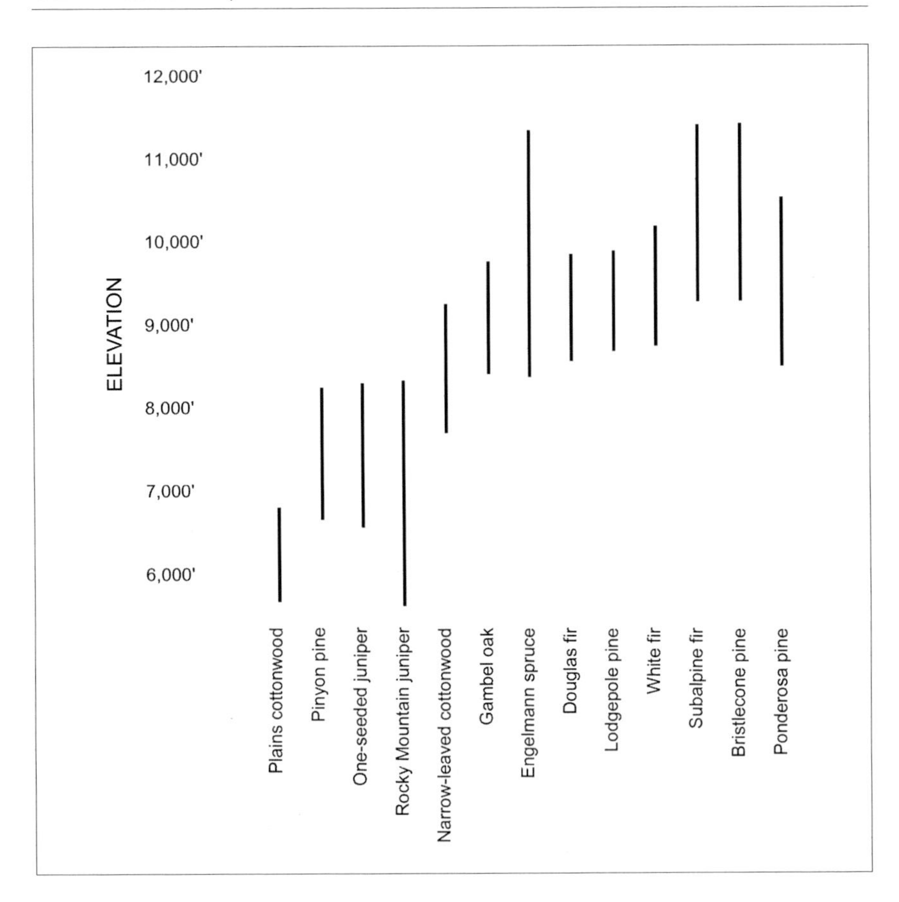

completed. Especially in mountains, the imbalances are permanent because of differential heating of slopes and diurnal (day-night) cycles of alternating heat and cold. The winds in a valley with a western exposure may be quite different in direction and intensity from those in an adjacent valley with an eastern exposure. Such local winds are called *mesoscale circulations*.

Normally, the general stirring of the atmosphere through local winds prevents steep gradients from developing and weather changes are generally gradual. Violent weather occurs when very steep pressure gradients develop. This is due to continental-scale (macroscale) high- and low-pressure areas that are superimposed over local heating and cooling patterns by continent-wide jet streams. The geographic ranges of species are usually determined by climatic extremes, conditions that may occur quite infrequently.

Finally, and very important for living things, are boundary layers, areas at the interface between smooth, unobstructed wind and the earth's surface.

They add microscale complexity that greatly affects the distribution of living things. Tree trunks emerging from a forest canopy, rocky outcrops, and buildings all add roughness that forms eddies, areas of wind speed-up and slow-down. Within boundary areas, microclimates develop. Tree bark, leaves, and ground surfaces all provide great variation in solar and wind exposure.

Wind desiccates living organisms, and species vary greatly in their tolerance to it. The survival of some organisms may amount to a matter of placement that varies within inches. Many lichens, for instance, may survive only on the leeward side of a tree, the windward exposure being too harsh. This might be due to either winter or summer prevailing wind direction. Considering individual species, ecologists have a precise understanding of microclimate needs in relatively few cases.

4. *Moisture.* Partially determined by elevation and exposure, precipitation and the resulting soil moisture are highly uneven in the western mountains. When people at lower-elevation flatlands are complaining about parched conditions, hikers in a 10,000-foot valley may be enjoying a view of swollen streams that begin in long-persisting snow fields. Timing of precipitation is important. Deep, persistent, high-elevation snow fields are important reservoirs that slowly release their contents as soil moisture and stream flow. By contrast, summer rains, although welcome, have a much shorter-term effect. Rain shadows can influence exactly where rain falls. The windward side of a range will often enjoy twice the rainfall as the downwind (leeward) side. Finally, atmospheric moisture, or relative humidity, is low here, an average of 10% or less.

5. *Substrate.* Underlying geology (see Chapter 4, Geology) determines the nature of soil nutrition. The absence of certain chemical nutrients may severely restrict which species of plants can grow. Soil pH, a measure of acidity or alkalinity, determines the solubility of nutrients that plants must take up through their roots. In the Spanish Peaks region, plant life encounters few unusual nutrient conditions, but there are a few differences. Pinyon pines and junipers do best on limestone soils. Where they are absent, the soil is likely to be more acidic. Ponderosa pines here prefer quartz sandstone–derived soils and enough water. They develop well around here on ridges underlain by Dakota and Lyons sandstones.

Rich and fertile topsoil is a thoroughly integrated mixture composed of rock-derived soil particles mixed with a generous amount of partially decayed organic materials. Soil microbes, fungi and bacteria, are particularly important in digesting the organic materials and making them soluble. Such a soil provides a constant stream of nutrients that are needed by fast-growing plant roots.

This ideal usually does not describe soils of arid regions. The combination of low humidity, low rainfall, and temperature extremes, as well as a small organic materials reservoir, causes slow growth and slow microbial decomposition. Frequently, the soils are relatively thin and poorly developed. Once disturbed (e.g., by the feet of grazing animals), the soil is quite vulnerable to sheet erosion. On exposed slopes, vegetation is often distributed sparsely, since there is insufficient water for numerous adjacent root systems. Under these climate conditions, irregularly occurring downpours cause immediate loss of the finer soil particles, often leaving an impoverished topsoil layer or removing it entirely.

BIOLOGICAL FACTORS

Diversity of species in the Spanish Peaks region is surprisingly high (see the checklists later in this book) considering the relatively low moisture availability and other environmental stresses. At the same time, the number of species in any given neighborhood is low, since each species has its own range of tolerance to such variables as moisture, low and high temperatures, and other aspects just discussed. Our elevation range magnifies the variation in these factors.

All species have life cycles, periods of establishment, growth, maturation, and reproduction. At each stage, individuals face different challenges. At this point we can consider some of the biological aspects that allow only some organisms to survive and thrive. For many individual species, we know little of the specific growth requirements and range limits, yet the factors that influence plant ranges are generally understood. They are as follows:

1. *Competition.* Nature's substrates bear little resemblance to your garden. Almost all soil space is already occupied with root systems of mature plants that are happily exploiting the local minerals and water. Most seedlings have little chance of surviving unless they happen to land in a place where another plant recently died. Also, we now understand that seed germination and seedling establishment, especially for trees, require more restricted and specialized conditions than are required for mature plant survival. For many species, these optimal environmental conditions are infrequent, occurring only every several years. Optimal conditions include prolonged adequate moisture or, as is the case with lodgepole pine and aspen, germination after infrequent fires.

2. *Pollination.* Seeds can only be formed if an ovary-containing flower receives pollen from the same species so that embryo fertilization can take place. Conifers and many flowering trees are wind pollinated, but the major-

ity of flowering plant shrubs and herbs require flying insects to carry pollen between flowers.

3. *Dispersal.* Plants have fruits containing seeds that must reach a favorable place where they can become established seedlings. Although some seeds or fruits are adapted to be easily wind carried for long distances, many others are animal dispersed. They must be eaten and passed through the digestive tract unharmed or be carried on the surface and dropped in a favorable location. The dispersal range for animal-carried species is obviously more limited than with wind dispersal. There must be a parent plant in the neighborhood, a local history, or a presence.

When you reflect on it, it is a marvel that a healthy plant survives in nature where you see it blooming and growing through photosynthesis. It is the winner of several sweepstakes on the path to survival. To begin with, millions of pollen grains and thousands of ovules are produced for every successful fertilized embryo. At least thousands of seeds are produced for every one that lands on a suitable germination substrate. Thousands of seedlings attempt to establish themselves for every one that finds a suitable space or experiences just the right weather conditions that year. Finally, immovable plants are vulnerable to the occasional fire and all manner of predation from worms and beetles to the larger grazing animals. We must conclude that all successful plants deserve a lot more respect than they often get.

4. *Plant defenses.* Even though plants appear to be sitting ducks for all manner of environmental insults, they are not as flimsy as they appear. Over the millennia, differential survival has selected for a variety of chemical and physical defenses that serves them well in the battle for energy and resources. In addition to the presence of thorns in some species, most plants are virtual chemical factories. They have a diverse "secondary metabolism" that produces bitter and toxic compounds that discourage both grazing animals of all sizes and other pest infestations. Over millennia, humanity has tapped into this discovery for flavorings and medicines of many kinds.

Consider a recently studied western example. One of life's pleasant and satisfying sensations often associated with coming to higher elevations is the odor of conifers, the "sweet pine-scented air." As consumers, we find pine scent so inviting that pine-scented cleaners, closet deodorizers, and such are big sellers everywhere. In nature, "piney" odors are due to monoterpene emissions from conifer needles, and they are a significant source of hydrocarbons in the inner atmosphere where clouds form and our weather occurs (see Jung et al, 2001). Although found in all of North America's coniferous habitats, piney odors are especially noticeable at higher elevations and at low humidity, two conditions that accelerate their evaporation from needle leaves.

In the vicinity of ponderosa pine needles, for instance, at least eight separate volatile monoterpenes contribute to this characteristic odor. The compounds vary in proportions among the various species of conifers and, to some extent, genetic variation within species. There are also significant differences between winter and summer emissions, because they are more volatile at higher temperatures.

When the food-conducting phloem tissue in the bark is wounded, the wood is exposed and begins to bleed "pitch" or oleoresins. This is the hard-to-remove goo deposited on your hands and clothes when you carry home the family Christmas tree. These oleoresin polymers are significantly high in monoterpenes.

Now we add the element of plant-animal interactions. Consider the winter food habits of Abert's squirrel, resident in ponderosa pine forests just west of here. When cold weather sets in, the squirrels regard the pine's inner bark as choice winter food. As we would imagine, large numbers of squirrels eating inner bark in some areas result in severe defoliation of these noble pines (see Snyder, 1992).

Keen observers have long noted that some individual pines are not targeted by the squirrels. Recent air sampling studies of non-eaten pines have shown them to have a significant genetically based difference in monoterpene emissions compared with the squirrels' preferred food trees. Biologists would expect that such selective foraging on these pines will eventually cause differential survival of tree genes for monoterpenes; that is, the survival of a greater proportion of "squirrel-resistant" trees.

PLANT COMMUNITIES OF SOUTHERN COLORADO

In any given location, a community or association of plants and animals is present that shares similar requirements. Communities are usually named for one or several species that appear to dominate the landscape at some elevation and moisture availability. Between these communities are boundary areas, such as between forest cover and grassland. These are called *ecotones*, and they often harbor a distinctive flora and fauna. What follows is an introduction to the basic communities and their characteristics.

Initially, we will follow an elevation-based classification with approximate elevation limits given. The actual elevation limits of species are strongly influenced by exposure direction and other microclimate factors as described above.

GREAT PLAINS GRASSLANDS (6000 Feet and Below)

Our position at the edge of the mountains is the western limit of the Great Plains. This huge ecosystem, occupying parts of 10 states and 450 counties, is named for its preponderance of grass cover. The lack of trees over most of the Great Plains is due to low amounts of moisture (10 to 20 inches per year) and, worse, very uneven precipitation within and between years. Our portion of the range can be classified as a "short-grass steppe" where the species are mostly deep rooted and drought resistant. The mixture also includes broad-leaved herbs, called *forbs* by range managers, which are dominated by legumes and members of the aster family. The area is historically important for the grain framing and ranching attempts of settlers and their descendants. Productivity is notoriously low and unreliable, a circumstance that has led to much rural poverty.

LOWER ECOTONE (6000 to 7000 Feet)

This region at the edge of the uplands includes the lower timberline where Great Plains grassland is gradually replaced by woody vegetation. The border may be sharp or vague with grassland and trees interfingering within this elevation. In northern Colorado, the woody plants are various shrubby species. Here, the dominant woody plants are pinyon pines and junipers, usually referred to as the "P-J" community. At the upper limit, P-J grades into ponderosa pines. Ecotone appearance can also be influenced by periodic wildfires.

MONTANE ZONE (7000 to 9500 Feet)

Montane plants live on the sides of mountains; this habitat can be divided in many places into lower and upper montane zones. In this region there is less distinction. The lowermost communities here are the P-J communities, which grade into patches of Gambel oaks on more exposed slopes (Fig. 29), and ponderosa pines in moister areas. Aspen is found throughout this zone and up into the subalpine zone. In the upper montane zone, white fir and Engelmann spruce are added. Limber pine, often listed as a marker for this zone, actually has been found to have a very wide elevation range and is found as low as Pawnee Butte (5200 feet) on the plains to the east.

SUBALPINE ZONE (9500 to 11,500 Feet)

The subalpine fir is added here, but white fir and limber pine continue. Engelmann spruce and bristlecone pine persist to the alpine timberline. Frequently,

Fig. 29 (above): Hillside, looking northwest from Route 12 between Cuchara and La Veta. Shrub patches are clones of Gambel oak. Different coloration among clones shows up well in the fall.

Fig. 30 (left): View to the northeast from upper slopes of Trinchera Peak. Blue Lake is visible in the middle distance. Beyond the Cuchara Valley are the three White Peaks. Several vegetative zones are visible.

as at lower elevations, these conifer species often occur as monospecific stands. This zone is the highest continuously forested landscape (Fig. 30).

ALPINE/TIMBERLINE ZONE (11,000 to 12,000 Feet)

Subalpine species reach their range limits in this zone, where they are capable of surviving as stunted forms. Among all of our woody species, they alone can withstand a very short growing season and extreme weather. The subalpine species drop out as growing conditions deteriorate locally, leaving only two conifer species. On very exposed ridges Engelmann spruce survives with a dwarfed horizontal trunk that in some parts of Colorado can grow slowly and laterally across the landscape. They die off on the windward side while producing new growth on the leeward side. Gnarled bristlecone pines may survive for centuries with most of their bark killed and blown off. Only a strip of living cambium remains to supply nutrients to perhaps one remaining branch. Other woody and herbaceous plants are quite reduced in stature, becoming cushion plants. Slow-growing willow species, old for their size, develop a low and compact growth form. This zone ranges upward to the environmental limit of growth, finally ending at bare rock.

It is worth noting that high-elevation rock piles such as the summits of the Spanish Peaks have a thriving community of spiders, flies, and perhaps a half-dozen other tiny animals. There appears to be nothing to eat except each other, a circumstance that would make an interesting study.

ZONES AND COMMUNITIES THAT CUT ACROSS ELEVATION BOUNDARIES

RIPARIAN WOODLANDS AND SHRUBLANDS (5000 to 10,000 Feet)

Riparian communities comprise plant and animal species concentrated along rivers and streams and their adjacent valleys. The root systems of most of them are in contact with the water table for much of the year.

From the Great Plains westward, the most abundant plant growth is found along watercourses. At a distance, their presence can be detected by the rows of woody shrubs and trees winding through the surrounding dry and grassy plains. A characteristic feature of this habitat is periodic flooding, and only a few woody plant species of the North American flora have root systems that are adapted to more than brief submergence.

Riparian woodland plants are broad leaved and deciduous. Most typical are species of cottonwoods, the tallest broad-leaved trees of our region, and the related willows of various species and sizes. Willows range in growth habit from small shrubs to larger trees (Fig. 31). Box elder, a species of maple, is also common. Riparian zones are also noted as islands of animal diversity in the West. It has been estimated that the majority (90%) of mammals and birds living in the region are dependent on this habitat. Healthy running surface water contains numerous organisms that are the basis of a complex aquatic and terrestrial animal food chain (Fig. 32).

For obvious reasons, most riparian habitats have been subjected to intensive human use over the last 150 years, resulting in degrees of degradation all the way to total destruction. Stream banks are often severely damaged by the feet of grazing animals or by cropping, mowing, installing irrigation features, and building dams. Stream use can be properly managed, and restoration of vegetation is underway in many parts of the West. Frequently, degraded stream banks are invaded by non-native weedy plants including tamarisk (salt cedar), Russian olive, bromegrass, Canada thistle, and leafy spurge.

Tamarisk

Of the above, tamarisk, which some call "the plant from hell," is the most noxious. Introduced to the West in the 19th century, this pretty shrub with feathery green branches that bear small pink flowers has quickly spread across the Southwest over disturbed riparian habitats. A mature plant can produce more than a half-million tiny airborne seeds that are capable of germinating immediately. Water managers know it as a premier phreatophyte. When a large plant's roots reach the water table, it can transpire about 300 gallons of water per day into the atmosphere. Colonies of tamarisks on stream banks have caused some streams to dry up. Also, their roots accumulate salts that are transported up the stems and deposited in the leaves. When leaves are shed and decay, salt concentrations increase on the soil surface. Salty ground prevents seedling establishment of salt-sensitive native species, thereby giving tamarisk a competitive advantage. Finally, tamarisk has little nutrient value and is almost worthless to wildlife. Native bird diversity in tamarisk communities sinks to about 1% of normal levels when compared with native shrub populations.

SOIL CRUST

Here, as in much of the arid West, the ground's surface on lightly trod areas is covered with a sort of water-absorbing sponge. The texture of crumbly toast

Fig. 31: The Cuchara River descends Bear Lake Canyon, with its banks lined by Bebb's willows.

Fig. 32: The Purgatoire River, looking east along Route 12 toward Fisher's Peak. Rivers and their valleys are home to most of the region's biodiversity.

when dry, this half-inch-thick layer appears covered with what looks like dead lichens and mosses (Fig. 33). It is where living and physical aspects of substrate thoroughly intertwine in an unusual way. Ecologists know it as a microscopic living community called cryptobiotic crust (meaning "hidden or microscopic life"). Studies of these crusts from several areas of the West have been published, although not for this area. Unfortunately, the huge ecological significance of the cryptobiotic crust is hardly recognized or appreciated by people who attempt to wrest a living from arid landscapes.

While up on Cordova Pass at about 11,000 feet one afternoon, I collected a dime-sized piece of crust from an ungrazed open area. I placed it in a culture dish with some water, and within seconds its volume expanded about 10-fold. While dissecting it under a microscope, I found a dense community of a couple dozen species of living things. They were classified among four different algal orders and three different fungal orders, plus a tiny moss. Soil particles of various size classes were intermixed. The fungal strands and moss rhizoids formed an entanglement that bound the whole community together. Studies of similar crusts from Utah have yielded 45 to 60 species from three different habitats.

Fig. 33: Cryptobiotic crust on the soil surface at about 9500-foot elevation.

Biologists know that cryptobiotic crusts, when present, are one of the best indicators of soil health. In good condition, these crusts absorb huge amounts of water, preventing almost all runoff from rain events. This greatly reduces regional flooding because excess water gradually seeps into the topsoil. Especially important in arid regions, intact crusts also greatly slow the process of surface evaporation. In grassy areas, the photosynthetic productivity of crust algae has been found to equal that of the grassland plant species. The crust's cyanobacteria absorb significant amounts of nitrogen and make them available for root absorption. As a final virtue, crusts do not carry fire well.

The bad news is that they are easily destroyed by heavy feet (yes, cows). Once destroyed, crusts may take 100 years to return to a mature condition. Given their outsized contribution to landscape productivity, there should be more research on how to reestablish crust organisms where they are lost, as well as more research on new ways to harvest net energy from the land in a sustainable way that takes advantage of this new understanding. It seems that the future of making a living on arid lands will require a better understanding of climate/soil interactions and the development of agricultural systems that are best adapted to these conditions.

There is still much to be learned. The interaction between plant roots and the soil to which they attach is complex and not easily interpreted. Root health is influenced by acidity, proportion of organic materials, soil particle size, mineral nutrient availability, aeration space, concentrations of atmospheric gases, and growing season length. Plant species vary in their tolerance for these factors, which are quite important in determining their range limits. Soil factors take some instrumentation to measure but have a major impact on most species.

TREE PESTS OF THE REGION

Whether you are hiking or driving in the conifer zone, you will notice extensive patches of dead trees in the landscape that are caused by a variety of small parasitic organisms. The patchy outbreaks are due to the monocultural growing conditions of this type of ecosystem, as well as environmental stress. But let's consider why, in some years, pests seem to take a heavy toll on plants. For instance, why did budworms wreak havoc on the new growth of our spruces and firs in the summer of 2002? We will consider the life histories of three locally important organisms, the mountain pine beetle, the western spruce budworm, and the Ips beetle.

MOUNTAIN PINE BEETLE (MPB) (*DENDROCTONUS PONDEROSAE*)

This species attacks pines, while its relative *Dendroctonus pseudotsugae* attacks spruces and firs, especially Douglas fir. The beetles are small and brown and about ⅛ to ⅓ inch long, and exact identification is best done by a specialist. For the MPB, the preferred hosts are ponderosa, lodgepole, limber, and, to a lesser extent, bristlecone and pinyon pines. Although the beetles are always present at low population levels, in certain years, population outbreaks occur that are very damaging and extensive.

Life History

Adults leave the bark of a tree in late summer or when the tree crown turns red due to dying. Females fly away on a quest for green trees, often in a coordinated mass attack. Mating occurs, followed by boring a vertical gallery through the bark into the cambium, the food-rich growing layer. About 75 eggs are laid, which soon hatch. The larvae move away from the egg gallery, making new lateral galleries (tunnels) that separate and kill the food-transporting system in the tree. Larvae overwinter in the bark, pupae form in June and July, and then new beetles emerge in late summer through September.

The major outbreaks ravage the landscape infrequently but are mostly related to drought stress. Other factors that facilitate outbreaks are anything that makes tree growth less vigorous, including tree age, tree crowding, poor nutrient conditions, mechanical damage, or fire damage.

Among MPB symptoms that specialists look for are (1) popcorn-shaped pitch tubes on the bark at the start of tunneling, (2) boring dust at the base of the tree, and (3) the crown turning yellow or reddish. For cut trees, the sapwood will often be stained blue. This is due to a fungus species carried by the beetles that also infects the tree.

On healthy trees, beetles are often overwhelmed by pitch flows at the beginning of tunneling. Weakened trees are ill equipped to fend off such an invasion.

Control

There is no good weapon. Some recommend thinning stands, burning infected trees before beetles emerge, or wrapping a cut tree trunk in plastic and giving it a solar heat treatment to raise the temperature to 110°F or hotter. Some pesticides are used but are not a good idea. There are many predator organisms in a healthy ecosystem, such as other insects and birds, which can keep MPB populations in check if given a chance. Watering yard trees gives them strength.

WESTERN SPRUCE BUDWORM (WSB) (*CHORISTONEURA OCCIDENTALIS*)

This is the most important tree defoliator in the West. Douglas fir is the favored host in Colorado, but WSB also will colonize white fir, Engelmann spruce, blue spruce, and subalpine fir. The larvae eat the new growth of branch tips of host trees (Fig. 34). (Another budworm of ponderosa and lodgepole pines, *Choristoneura lambertiana*, has a similar life cycle.) The adult is a small, mottled, rusty brown moth, ranging from tan to almost black. It is about 1 inch long. In Colorado, the WSB is present from June to early August.

Life History

After mating in summer, the female moth places the overlapping masses of 25 to 40 green eggs on the underside of needles. The larvae hatch in 10 days and, without eating, move to bark scales or lichens to remain dormant through the winter in silken hibernaculae. In April through May, the larvae migrate to foli-

Fig. 34: Engelmann spruce (*Picea engelmanii*) branch. Budworm damage is present on branch tips.

age where they may mine old needles. Within two weeks, they enter and eat developing buds and grow quickly. They web the needles together to form a protective chamber. Late larvae have brownish olive bodies and brown heads. By the end of June they pupate, and adult moths emerge within one week.

The larvae may eat all new foliage, which reduces growth significantly and causes the twigs to die. In long-running outbreaks of three to five years, one in four trees will die. Bark beetles often help finish off the survivors. In large outbreaks, the budworms can defoliate up to 2 million acres, although most infested areas are far smaller.

Control

In a normal healthy ecosystem, budworms have many predators and parasites and are hurt by bad weather such as late spring freezes. Thinning is the long-term solution. Promote yard tree vigor by watering. Chemicals work but also attack the budworm's predators.

THE IPS BEETLE (*IPS CONFUSUS*)

This organism is one of many species of *Ips* that attack a large variety of conifer species in North America. Our local species has caused alarming damage to many of our oldest pinyon pines, slow-growing trees that may take a century to regenerate in this area. Again, drought stress makes the trees vulnerable. The beetles are attracted to stressed trees that cannot outgrow the infestation as they can when normal soil water is present.

Life History

In summer, the female beetle lays eggs under the bark of healthy trees. The hatched larvae tunnel beneath the bark, eating the soft and nutritious cambium and phloem areas. When the cambium is entirely girdled, the tree can no longer transport nutrients to the roots and the tree dies. The beetles overwinter. They emerge from bark in early spring and locate other trees to attack.

Control

Insecticide sprays are valuable in protection of remaining healthy trees. Information is available from the Colorado State University (CSU) Cooperative Extension agent or the Forest Service.

WESTERN TENT CATERPILLAR (WTC) (*MALACOSOMA CALIFORNICUM*)

Among broad-leaved woody species, this tent caterpillar is the most common locally among a number of species of webworms found in Colorado. During their outbreaks, the silken tents that they build in tree branch crotches can be seen from even a casual glance at the landscape. Although their infestations of aspen and mountain mahogany are especially conspicuous, they are also found to a lesser degree in other tree species. The caterpillars are colorful and up to 2 inches, while the light tan moths are about ¾ inch long (Fig. 35).

Life History

In late summer, fertilized female moths lay egg masses that adhere to twigs of the host trees or shrubs. They overwinter in this phase until early spring. At that time, before bud break on the host plant, small caterpillars hatch and begin building silky tents in the crotches of branches. These tents give them shelter and a resting place during the day. The tents expand as the caterpillars themselves enlarge. For weeks, they chew away on emerging leaves, often removing all from any branch within reach. They may also feed in daylight, although they are vulnerable to predation from birds, other insects, and virus and bacterial diseases.

Fig. 35: A western tent caterpillar swarm on the end of a cherry tree branch.

In late spring, when fully grown, they wander away from the tent, spin a silk cocoon, and pupate. After two weeks, they emerge as light tan moths, fly off, and mate. Later, females lay an egg mass on the twig of a host tree. A single new generation is produced each year.

It takes a number of years for population densities to have devastating effects on host tree populations. I have seen them totally defoliate an aspen population of more than 1 square mile in extent. Forest aspen trees, being attached by underground root connections, can withstand several years of defoliation. However, if other stress factors occur, such as drought or unusual winters, the trees definitely have their lives cut short, leading to ghost forests. Because of the tent caterpillar's various vulnerabilities, successive large annual outbreaks are rare.

Control

A number of pesticides will work (see the CSU extension website in Resources) as will pruning. A common approach, the burning of tents with some sort of torch, usually does more damage than the caterpillars. In yards, the tents can often be removed manually. Otherwise healthy trees in well-watered places can usually withstand one or two outbreaks without long-term damage.

FOREST SUCCESSION

Forested habitat is the best place to see secondary succession. This is the long-standing observation that species composition changes between the times that trees colonize a bare area and when a mature forest is found on that site. This is mostly due to two factors: variations in shade tolerance and variations in seed production capacity. Understanding succession lets you estimate the age of a forest since the last disturbance.

PIONEERS

In the montane zone, in the aftermath of fire, two tree species show up very early. The first conifer species, the sun-loving lodgepole pine, readily colonizes freshly burned areas (see Chapter 10, Conifers). Therefore, a tract of these trees is likely to be all the same age. Lodgepole pines, however, do not grow well in shade, so the understory specimens in such a forest are spruces and firs that do grow well in the shade of pines. The lodgepoles are relatively short lived and often begin to die within a century. Spruces and firs, capable of living much longer, gradually replace them. A mature stand of lodgepoles and their understory can be seen on the Dodgeton Trail near its intersection with the Indian Creek Trail, and at Uptop on Old La Veta Pass.

The second pioneer, and the most common local broad-leaved tree, is the trembling aspen (see Chapter 11, Woody Flowering Plants). Aspen stands may also be short lived and gradually replaced by the slower-growing firs and spruces (Fig. 36). The species seeds more often after logging but occasionally after fire. Good examples of aspen stands can be seen on many trails.

The south end of the Dike Trail, upper part, shows many dead and dying aspens being replaced by a vigorous crop of conifers that used the shade of the aspens as a nursery. In this area, lodgepole pines and aspens have grown about 120 years since the last disturbance, about the limit of these species before their populations become senescent.

Fig. 36: Aspen forest with understory of Engelmann spruce, at the Farley Wildflower Overlook one-half mile east of Cuchara Pass.

As a side note, there are places in the Rockies, perhaps in this region, where dense aspen stands persist. The species has the capacity to produce new stems (trunks) from root sprouts, which means that a single massive root system may be connected beneath an entire stand. Some populations in the central Front Range are considered to be thousands of years old, having persisted from the Late Pleistocene age. I have seen no evidence of such longevity in this area.

A STORY OF PHOTOSYNTHESIS, OR, WHY DO EVERGREENS DOMINATE THE FLORA OF AREAS WITH SEVERE CLIMATES?

As a general rule, when you find broad-leaved deciduous and needle-leaved evergreens in mixed populations, the mean temperature of the coldest month

(January) is about 1°C (34°F). Colder sites tend to have a dominant conifer population (see Spruegel, 1989).

Consider photosynthesis, the source of all metabolic energy on earth. Here, trees that belong to flowering plant families all have deciduous broad leaves that are well adapted as "flat plate collectors" for intercepting solar energy. When struck by visible light, the unstable chlorophyll molecules in leaf cells transfer energy in a way that allows atmospheric carbon dioxide to be chemically changed into the sugars and starches. These are the metabolic fuels that make plants grow.

When you compare broad leaves, such as those of oaks or cottonwoods, with needle-like leaves of pines or firs, you would think that broad leaves would be much more productive of photosynthesis than tiny needles with little surface area. However, there are several counter-intuitive reasons why conifer trees are somewhat more productive and best adapted to cold, severe climates. First, deciduous leaves are "throw-away" organs that are shed after one year and operate only during the frost-free season. Conifer needles are retained on branches for several or many years.

Second, partly because of needle retention, and tighter packing, conifer canopies are deeper with leaves. In fact, when botanists have compared the crown or canopy of a conifer with that of a deciduous tree, the conifer was shown to have between two and five times as much exposed leaf area as the deciduous tree.

Third, the orientation of broad leaves to the sun strongly influences their photosynthetic efficiency. For full carbohydrate production, a broad leaf has to be within 15 degrees of perpendicular to the sun's rays. This means that as the sun tracks the sky on a seasonal or daily basis, broad leaves are seldom optimally oriented, and they may even be shaded. Frequently, their average photosynthetic production is just a fraction of their potential.

On the other hand, conifer needles are very productive over much wider angles of orientation to the sun. With their "hemispheric" or "bottle-brush" display, needle shading is less of a problem. Because of their numbers and orientation, daily sugar production is often at a higher overall level.

For conifers, there are many days during the deciduous tree dormant season when it is warm enough for photosynthesis, especially later in fall and earlier in spring. The capacity of needles to function at somewhat lower temperatures adds to their net efficiency.

Color is not accidental either. Studies in the Arctic have shown that dark green conifer canopies are exceptionally effective in absorbing heat. As wind causes snow shedding, the air temperature around conifer canopies is several degrees above that of adjacent all-white areas.

So we have a case of hare and tortoise. On ideally bright, warm days, broad-leaved canopies may exceed the photosynthetic production of conifer-needled canopies. But over an entire year, conifers are successful competitors, matching or exceeding the productivity and growth of deciduous species in the region.

TIMBERLINE

In the mountainous western United States, forest species survival is strongly influenced by elevation and exposure, and abrupt rather than gradual species boundaries are common. Considered from a distance, tree vegetation is found in a band with a lower timberline as well as an upper timberline. These vegetation boundaries look razor sharp when seen across a valley, but the transition is more complex when seen up close. The lower or dry treeline is where the grassy Great Plains are replaced by pinyon-juniper, sagebrush, Gambel oaks, and ponderosa pines. As implied by the name, moisture is the limiting factor for tree life at the lower treeline. Much of this area is rangeland and not accessible to the public.

At the upper or cold timberline, subtle changes take place that are reflected in unique species adaptations. Also, climbing, hiking, and fishing are popular in that zone, so I will describe it in more detail. Unless otherwise specified, my use of "timberline" or "treeline" will refer to cold timberline (Fig. 37).

When ascending a trail that goes through the forest to timberline, you notice that tall trees species gradually become dwarfed, bent, twisted, and gnarled and are finally lower than the height of the average person. These trees are often quite picturesque. At the extreme upslope boundary, woody plants are quite close to the ground and cushion-like. Often the stunted tree crowns are "flag form," where the branches extend only to the leeward side due to death of those directly blasted by strong, prevailing, and desiccating winds.

As originally described by German ecologists, the usual international term for these trees is ***krummholz***, which means "crooked wood." The deformities are slight at the margin of taller trees and gradually get more severe upslope. The krummholz line is often quite irregular due to factors of climate and local topography. Around here, the conifers that can make it to this elevation are bristlecone pine and Engelmann spruce. Subalpine fir may almost reach this zone, as found on the trail to West Spanish Peak.

The Big Picture

This transition is not always at the same elevation, although locally the average

Fig. 37: Looking northeast on the trail from Blue Lake to Trinchera Basin. An abrupt timberline transition is seen here, with willows and stunted spruces in the foreground.

is 11,500 to 12,000 feet. To the south (toward the equator) cold timberlines get higher; to the north (in the Arctic) they get lower until trees reach a final transition to tundra on a low plain. This phenomenon was first noticed in the early 1800s by President Jefferson's friend, Alexander von Humboldt, the great German founder of biogeography. At mid-latitudes, the tendency is for timberline to become lower by 500 feet per 100 miles northward, or 360 feet per degree of latitude.

Temperature

As you hike or drive upward in the mountains, you usually notice that it gets cooler. This general drop in temperature with elevation, other things being equal, is called the *lapse rate*. For this region, the lapse rate predicts a drop in temperature of 3.3°F per 1000-foot rise in elevation. Therefore, we know that high-elevation environments are almost always cooler. Even more interesting was the discovery of the 50°F isotherm. An isotherm is an average temperature line, resembling a contour line, that can be drawn across the landscape for a given month. In this case, the 50°F isotherm represents the average daily temperature for July, the warmest month. It happens that the 50°F isotherm matches the tree limit in the Northern Hemisphere. Colder than 50°F is higher than timberline, and warmer than 50°F is within well-developed forest. This approximation holds true for all temperate zone timberlines in North America and Europe.

Growing Season Length

The length of the growing season counts for a lot in extreme environments. At timberline, even hardy conifers need about 60 days of mostly frost-free conditions to thrive. Most species whose ranges do not extend into the higher-elevation forests are limited by hard frosts in late spring or early fall. The need for a frost-free growing season of a certain length is the principal limiting factor for the ranges of most alpine conifer species.

As is true of woody plants well below timberline, these plants are exceptionally frost hardy during the fall, winter, and spring but are frost sensitive during midsummer. Even mild summer season frosts, which happen almost every year, will damage new needles and shoots, a principal reason why such trees appear to grow so slowly. As you approach timberline, you can note the effects of growing-season frost damage on many of the younger branches.

Topography and Slope Direction

Timberlines vary by several hundred feet on different sides of a mountain peak or ridge. They are often higher on south-facing slopes because of the greater exposure to solar radiation and resultant heat buildup in the summer. Being out of direct exposure to the sun, north-facing slopes are often much cooler, but this is not an iron-clad rule.

If the prevailing winds are southwesterly, they may regularly sweep away the winter snows from the south-facing slopes, leaving tree seedlings unprotected and the site too dry. Then the north-facing slope will be moister and more hospitable to seedling establishment in spite of less annual warmth.

Gently rounded ridges at 11,500- to 12,000-foot elevation are common in the area. Their timberlines are highly irregular and can be called *wind timber formations* or *ribbon forests*. The ribbon forests are parallel bands of low-growing trees. The tree rows are separated by meadows of sedges or grasses. Being perpendicular to the prevailing wind, the trees act as snow fences that cause persistent snow drifts to accumulate in their lee. This prevents the growth of trees in those corridors. At the edge of trees, lower branches and shrubs such as Bebb's willow form extensive and dense, low mats. Such ribbon forests can be seen on the south side of Cordova Pass near West Spanish Peak, on the east slopes above Bear Lake, or above the North Fork of the Purgatory River. High ridges are particularly prone to developing harsh conditions because of year-round desiccating winds.

Concave alpine basins, often called *cirque basins*, were carved by the glaciers that once originated and moved within them. Their average summer temperatures remain lower because of cold air ponding, the falling and accumulation of colder air. Winter snow often accumulates to great depth in these basins. Otherwise, with no limitation on moisture and protection from wind, one might expect good tree growth, but temperatures remain too cool. Timberline may be as much as 2000 feet lower than on more exposed but warmer ridges. A good example of such a concave alpine basin can be seen at the top of the Bulls Eye Mine Trail on West Spanish Peak (Fig. 12).

The Historical Dimension

In some timberline environments covered by dwarfed living woody plants, you can see the carcasses of large trees standing as gnarled old masts, with many others lying prostrate in the meadow.* A careful inspection of many of them will show the results of fire scarring combined with sand blasting by the relentless year-round winds. The result is smoothly sculpted and bleached wood that is a pleasure to see.

Standing in this starkly beautiful landscape you might think: if this was once a forest, why hasn't it returned to forest? Answers have not been worked out for all areas, but four general factors seem to be at work. First is time. At these elevations where growing conditions are severe, it would take centuries for a forest to stand once again on this spot. Seedlings must begin their lives in the protection of a large rock, stump (Fig. 38), or log, where reflected heat is higher, wind is less, and soil water is adequate for at least several decades.

Second, it may be that a forest will not return at all due to climate change. The intensity of present year-round drying winds may entirely prevent tall trees from growing here in any foreseeable time period. Forests developed

*A good example is the east ridge above the headwaters of the North Fork Divide.

Fig. 38: Young limber pine growing in the lee of a dead trunk. Only with such protection can young trees survive in high-elevation, exposed habitats.

in another era may persist under a harsher climate and in the absence of fire. Trees already established and growing together have a moderating effect on air movement and humidity within the canopy; that is, they modify their own microclimate. If destroyed, microclimate conditions may have shifted too far from what it takes to establish a new forest.

Third is genetics. Timberlines represent the extreme limits of species survival. Here and, for that matter, anywhere at the extremes of a species tolerance, surviving trees are a naturally selected subset of the available genetic variation. It takes many generations for new genetic combinations to produce a sufficient quantity of seedlings to expand a species range. Disturbance may destroy local sources (seed banks) at the periphery.

Fourth, fires may destroy topsoil that took centuries to develop. Among the necessary topsoil components, the mycorrhizae may be missing. These are fungal partners that forest tree roots usually require for nutrient acquisition.

The ecologic details of survival in extreme environments remain a research frontier for most species. But these are tractable problems. All such environments can tell us much about climate change, historical events, past use, and abuse if we observe and learn to read the signs.

RECOMMENDED ACTIVITIES

In this chapter, I have attempted to lay out some of the biological and physical phenomena that occur in this beautiful landscape. Try keeping a daily natural history log of your property or favorite places in the region. Date your entries and note weather conditions. Observe what is happening as you take some of the trails that go through several plant associations. You don't need to name every species to do this. Just make a mental note of differences as you walk ever upward. Being outdoors is the most fun when you have some understanding of what is going on around you, especially when you see patterns and make discoveries through your increasing awareness.

CHAPTER 6

THE LIFE OF LARGE CHARISMATIC PREDATORS

WILDLIFE MANAGERS sometimes call charismatic predators and other large animals *flagship species*. They have the capacity to motivate the public to conserve habitat. A couple of species discussed here are probably extinct in Colorado but are included because of rumors and general interest. See Chapter 13, which lists all regional mammals, for further information about the habits of black bears and cougars.

BLACK BEARS (*URSUS AMERICANUS*)

Of late, bears are increasingly common around human settlements. Local hunters claim that the canceling of the spring bear hunt in the early 1990s by Colorado ballot initiative has contributed to bears losing a fear of humans. This, combined with early frosts that kill off the Gambel oak acorns and recurrent droughts that kill high country berries, explains why the bears become ravenous for food as they prepare for hibernation. To survive the winter, they need up to 25,000 calories per day from late summer onward. In addition, snowless "open" winters, where cold snaps of well below zero occur, make them vulnerable to dying in their dens. Weather is a variable that affects population size from year to year.

In the last several years, lowered fear of humans plus hunger have made bear-human confrontations and property damage more common throughout the region. Wildlife officers are called upon throughout the summer to move or destroy bears when their habituation to humans causes tragic consequences.

HOW DANGEROUS ARE THEY?

When cornered, teased, or provoked, black bears will attack people. However, statistics kept on black bear–human encounters show that bears only bother humans on the rarest of occasions. Sows will attack or bluff-charge if you blunder between them and their cubs. Avoid doing this by paying attention to what is happening around you as you travel. If you encounter a bear that has found something good to eat, do not approach it, as it will always defend a food supply. Back away slowly. Do not look directly into its eyes, as some bears regard this as a sign of aggression. If you linger in a productive berry patch, remain observant while you pick. Bears are fond of berries and use them as a major calorie source in their diet.

Herrero (2002), a leading authority on bear-human interactions, noted that 23 people have been killed by black bears from 1900 through 1980. On occasion (in Canada) they will treat humans as prey and attack without warning. Such bears are found mostly in remote areas and have no experience with humans. Bears have never killed anyone when suddenly confronted. To put this in perspective, the black bear population in North America has long hovered at about 600,000 to 700,000. There are literally tens of millions of human recreation days in bear habitat each year with only a tiny number of encounters.

Also, when assessing the dangers of the wild, consider this: According to the Centers for Disease Control and Prevention, dogs bite four million people per year in this country, leading to 300,000 emergency room visits. Fifteen to 20 people per year are killed by dogs. Collisions with deer kill more than 100 people per year. Cars kill at least 40,000 people per year. I conclude that being on trails in our forests is safer in most respects than "everyday life."

When setting off for tent camping in bear country, avoid taking perfumes or using perfumed deodorant. Bears are very curious and will investigate strong, unusual odors. Store your food in a bear bag (humorously called "bear piñatas" on the Appalachian Trail). Well away from your camp, hang the food bag on a line between trees too high for the bear to reach. This is at least 15 feet above ground and 6 feet away from a tree trunk. Keep a clean camp and keep food or smelly objects out of your tent. If you find an animal carcass when hiking, perhaps one day or several days old, and perhaps partially covered up, leave the area immediately. A bear probably "owns" it. It may not be far away and will aggressively defend it. Finally, when leaving your car or cabin, remember to secure doors and windows in your absence.

PROTECTION FROM ATTACK

In case of confrontation, some suggest carrying a gun. However, an effective gun is too heavy for most hikers to carry and more dangerous than helpful unless you are trained and use it frequently. A 38-mm riot gun that shoots rubber bullets has been shown to be very effective as a deterrent. Its use also requires training.

A better alternative is a mace canister that produces a spray containing capsaicin, or pepper spray. It is intended to temporarily sting the eyes of an aggressor. It works well, nearly always causing the bear to break off an attack. If you are interested in carrying mace, get the large can that sprays about 40 feet, which can be purchased in sporting goods stores. (And do not fire into the wind!) These larger canisters can be expensive (more than $40) but give peace of mind to some folks. Do not bother with the small mace canisters on key chains. They would be effective only if an animal is virtually on top of you.

COUGARS (*PUMA* [or *FELIS*] *CONCOLOR*)

The largest of American cats, cougars may be more than 8 feet, including the 32-inch tail, and weigh as much as 180 pounds. Because of their nocturnal activity and tawny coat color, they are rarely seen. The cougar's general hunting behavior is similar to that of other North American wildcats, except it can take larger prey. They are stealth hunters and masters of camouflage. They need cover or a ledge on which to hide. The cats prefer to surprise prey by a sudden bounding that can be as long as 45 feet and as high as 15 feet. They can jump down from ledges as high as 60 feet and silently land, running. A quick bite to the base of the skull dispatches a deer. More experienced cougars can take down a 1000-lb elk.

An estimate by those who keep such statistics for North America is that there have been 14 known fatal cougar attacks on people in the 20th century and 44 nonfatal attacks. Of these, half took place in the 1990s. As of this writing there have been two human deaths within Colorado, both in the past 10 years. One was a 10-year-old boy on a trail in Rocky Mountain National Park; the other a male student jogging near his high school in Idaho Springs. In California, two bicyclists were attacked on a popular trail and one was killed.

The commonly heard explanation for the increasing rate of confrontations is the increasing human use and development of cougar habitat. This is certainly true, but there has also been an explosion in the cougar population. Bounty hunting, which caused cougars to fear humans, ended in the United States and Canada in the 1960s. As of 1973, cougars have been protected by the Endangered Species Act. Since then, the cats have made a strong comeback. In some western states, a twofold to threefold increase since 1985 is estimated. One estimate suggests a population of at least 31,000 in 12 western states. This could be more than before human settlement, when cougars shared the top predator honors with large populations of wolves and grizzlies. In the Spanish Peaks region, there is no current census or estimate of cougar population size. However, they are being sighted more frequently around homes in the Cuchara Valley.

Predator populations gauge the health of the ecosystem. They can only increase on the basis of a large and healthy prey population. Although cougars feed opportunistically, deer are by far their favorite prey. The deer population has also exploded in the same interval due to management decisions by our game agencies. In addition, deer hunting by humans is said to be declining.

David Baron's excellent book, *The Beast in the Garden*, discusses in detail the reasons for cougar-human interactions. Through evolving environmental policies, we have placed cougars and people into closer contact than ever before. But this is a new set of circumstances and is not a return to a state of nature. The big cats are very intelligent and hungry and are adapting to the presence of humans, their pets, and their children.

Although I love wild animals, I have, with regret, come to a conclusion I wouldn't have anticipated earlier. At any opportunity, large cats should be run off by hosing them or throwing stones or sticks. The healthiest thing for the safety of both animals and humans is to have the cats fear humans. When large predators get comfortable around humans, the predators always pay with their lives.

PREVENTING COUGAR CONFRONTATION

Prevention:

1. At home or camp, keep your pets in at night and do not feed them outdoors at any time. Keep garbage secure.
2. If you are concerned for your safety, hike with one or more companions. Also note that dogs tend to be cougar magnets; very few are capable of taking on a cougar. Asked one game warden: Do you want a chased dog running to you for protection?
3. If you find a dead animal, a possible kill site, leave the area immediately.

In case of encounter or attack:

1. Do not act afraid. Be assertive. Do not turn your back, crouch, or run.
2. Do maintain eye contact. Look "big." Get the group together. Make loud, fierce shouts.
3. Use any weapon, knife, mace, walking stick, or rock, if you can get one without bending over.

4. Fight back. Target an eye with your thumb or weapon.
5. Report all cougar incidents to officials.

NOTES ON ANIMALS RUMORED TO BE PRESENT IN THE REGION

GRIZZLY BEAR (*URSUS ARCTOS*)

Some local folk swear they know someone who has seen a grizzly recently, especially in the La Veta Pass region. There has been no confirmed sighting in this part of the state for many decades. On occasion, very large "cinnamon" black bears are imposing looking and are mistaken for grizzlies.

For a long time it was believed that the last Colorado grizzly was killed by a federal trapper and guide from Monte Vista in 1952. Much later, wildlife biologists were surprised to confirm a report that the final "last" grizzly in Colorado was killed in 1979 by a bow hunter, reportedly in self-defense, in the San Juan Mountains about 90 miles to the west.

Since the early 1990s, there has been an intensified search for any remnant population in the San Juan wilderness area. This effort, led by the Colorado Grizzly Project, collaborating with Round River Conservation volunteers, has encountered numerous signs but nothing definitive. The National Biological Service of the U.S. Geological Survey also believes they may still occur in the San Juan Range, but we remain short of positive proof that a population still exists. At 158,790 acres, the huge, nearly impenetrable San Juan wilderness is adequate to support large bears. The Colorado Grizzly Project has proposed reintroducing them.

Grizzly bears vary greatly in weight. Females reach full size in early maturity at less than 300 pounds. Males continue growth each year of their life and often become twice as heavy as females. Six hundred pounds is a large male for the continental interior, but Alaska coastal males, whose diet includes salmon, may reach more than 1000 pounds and reach a length of seven feet. Never black, the bears are yellowish reddish brown or blond and have a humped shoulder and dished (concave) face. Tracks of their front feet show claw marks measuring about 4 inches long, but actual claws are about 1¾ inch long.

In addition to taking larger prey, grizzlies are not picky eaters. They will also eat carrion, fruits, roots, bulbs, fish, and adult and larval insects. In the Yukon, I watched one grazing on clover flowers. Grizzlies are known to exist in Montana and Wyoming, but their U.S. populations are small and they are on the federally threatened list.

GRAY WOLF, TIMBER WOLF (*CANIS LUPUS*)

The last wolf was killed in Colorado around 1940. As with grizzlies, "game inflation" still happens. A local rancher said he knows someone who is "certain he has seen wolves" following livestock, but this has never been documented. Large coyotes are also impressive looking but have a narrower face. The wolf's extermination was mostly due to an easily learned preference for killing livestock.

Wolves are much larger than coyotes, with large adults reaching a length of about 5 feet. They also stand taller and have a wider head. Studies are being undertaken currently in anticipation of reintroducing wolves in Colorado, but this has not yet occurred.

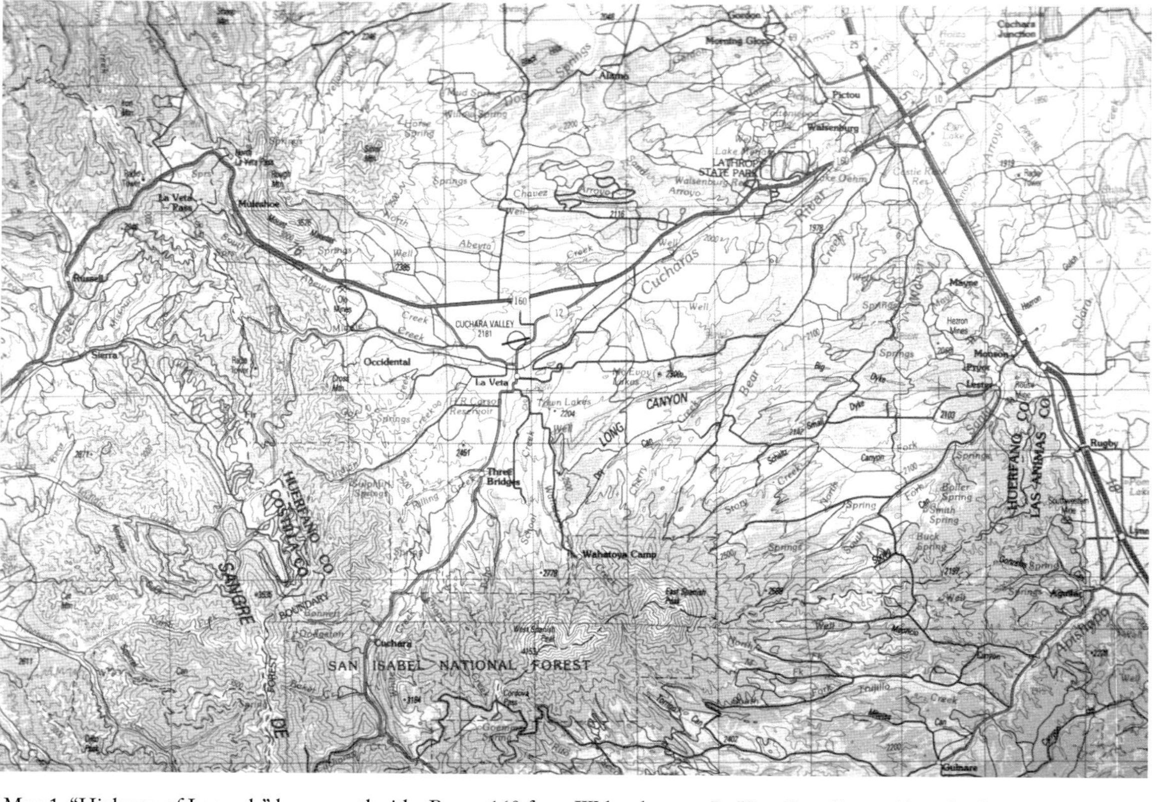

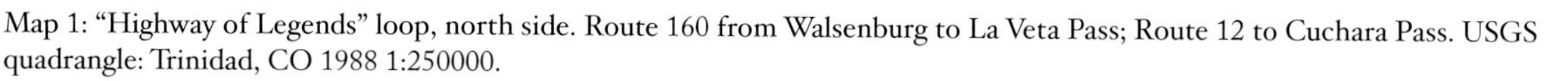
Map 1: "Highway of Legends" loop, north side. Route 160 from Walsenburg to La Veta Pass; Route 12 to Cuchara Pass. USGS quadrangle: Trinidad, CO 1988 1:250000.

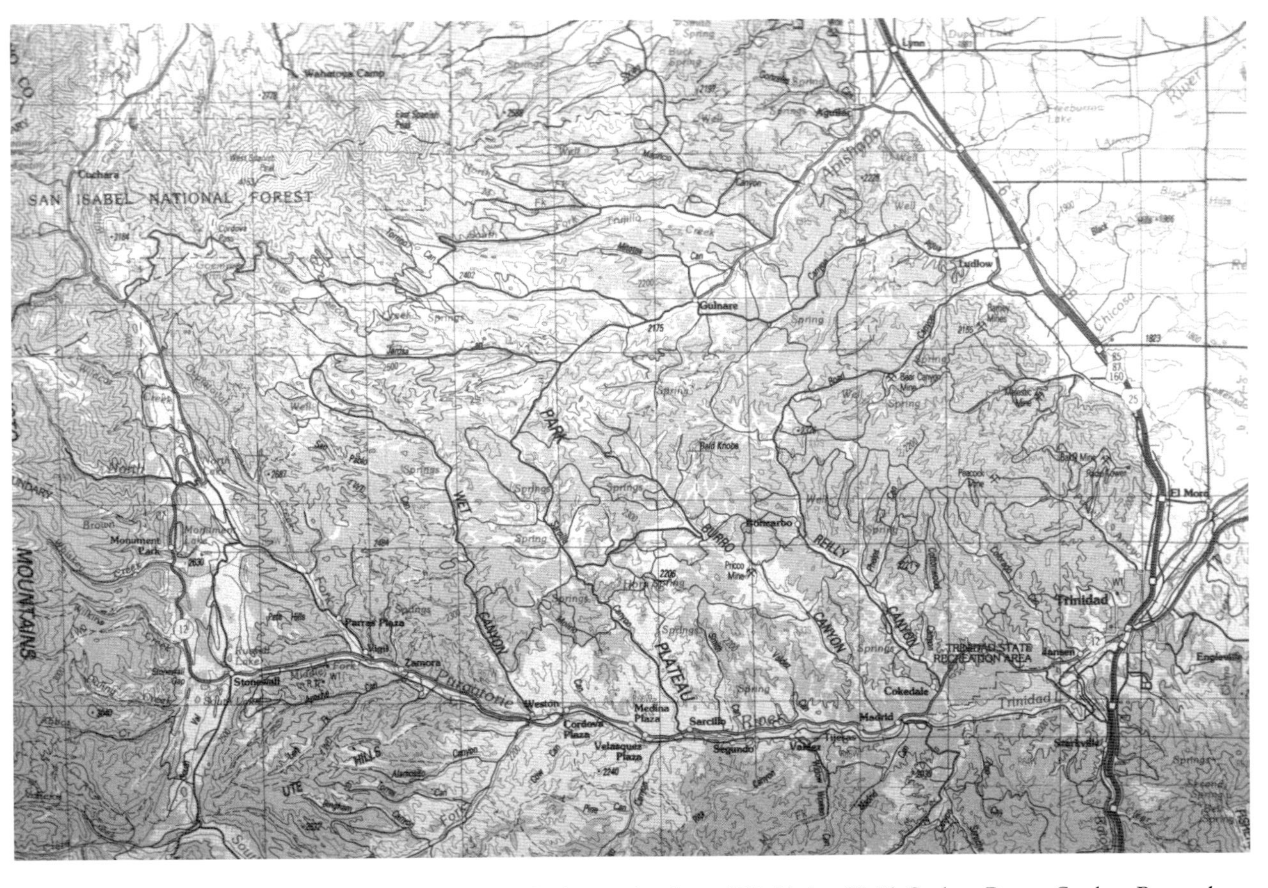

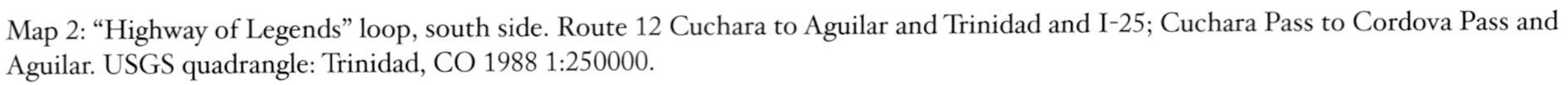
Map 2: "Highway of Legends" loop, south side. Route 12 Cuchara to Aguilar and Trinidad and I-25; Cuchara Pass to Cordova Pass and Aguilar. USGS quadrangle: Trinidad, CO 1988 1:250000.

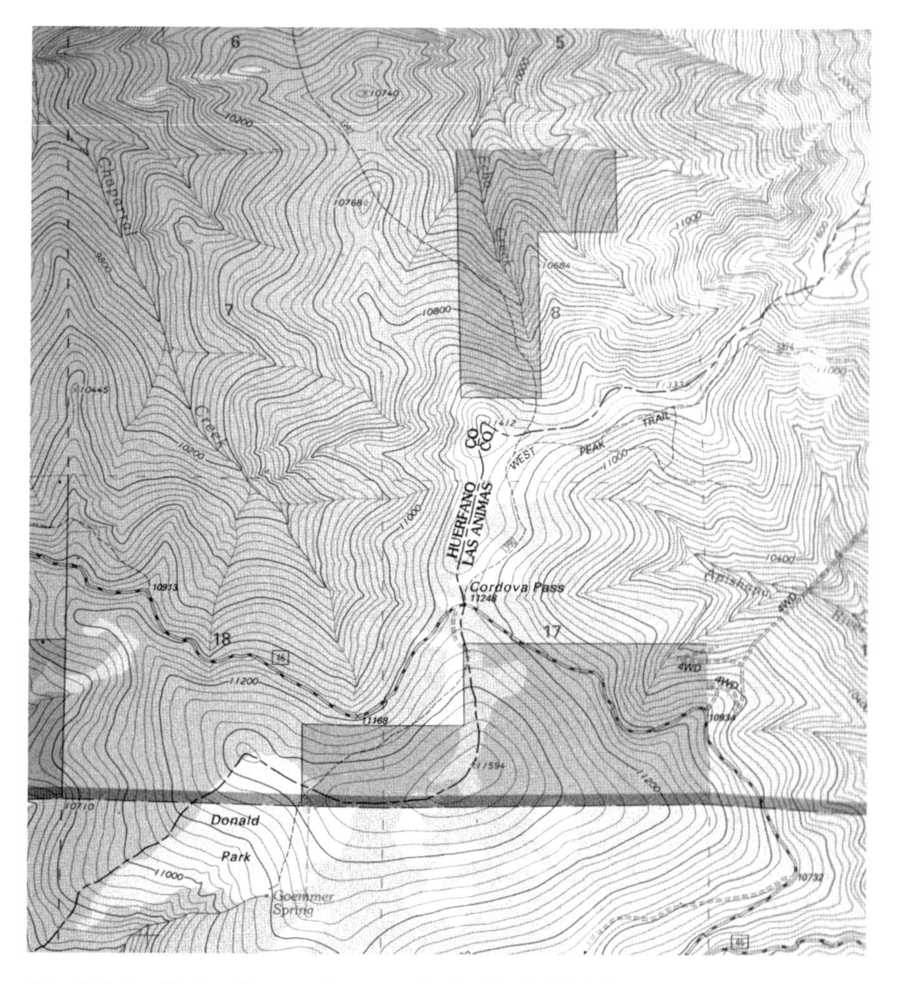

Map 3: Below Cordova Pass, on the west side, the Shaffer Trail descends northwest from a (barely safe) wide spot for parking. At Cordova Pass is the parking area and trail to West Peak, and a scenic spur to the west is marked. A fork on the northeast is the upper connection of the Apishapa Trail. USGS quadrangle: Cuchara Pass, CO 1994 1:24000.

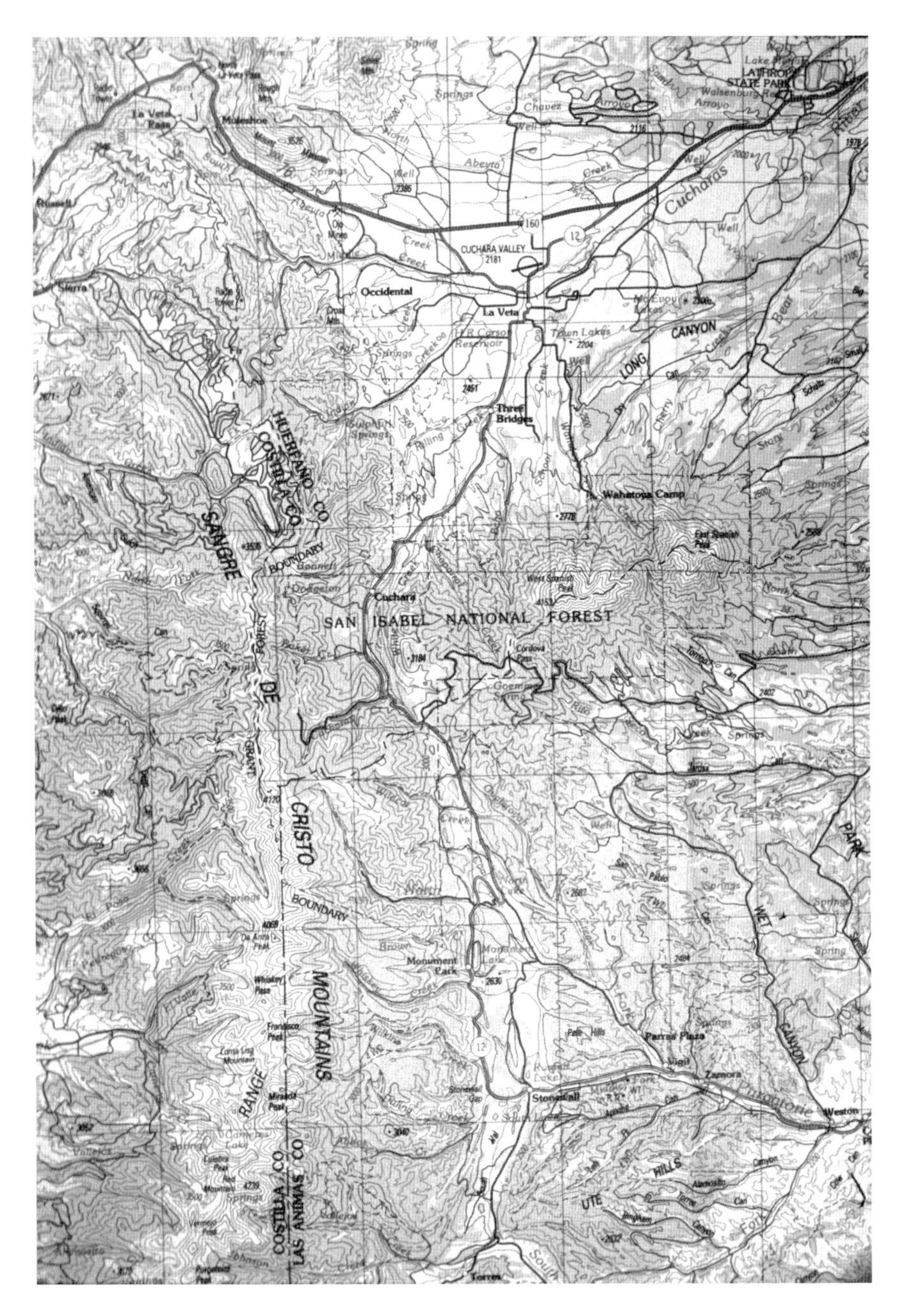

Map 4: La Veta to Stonewall; Lathrop State Park to the Spanish Peaks. USGS quadrangle: Trinidad, CO 1988 1:250000.

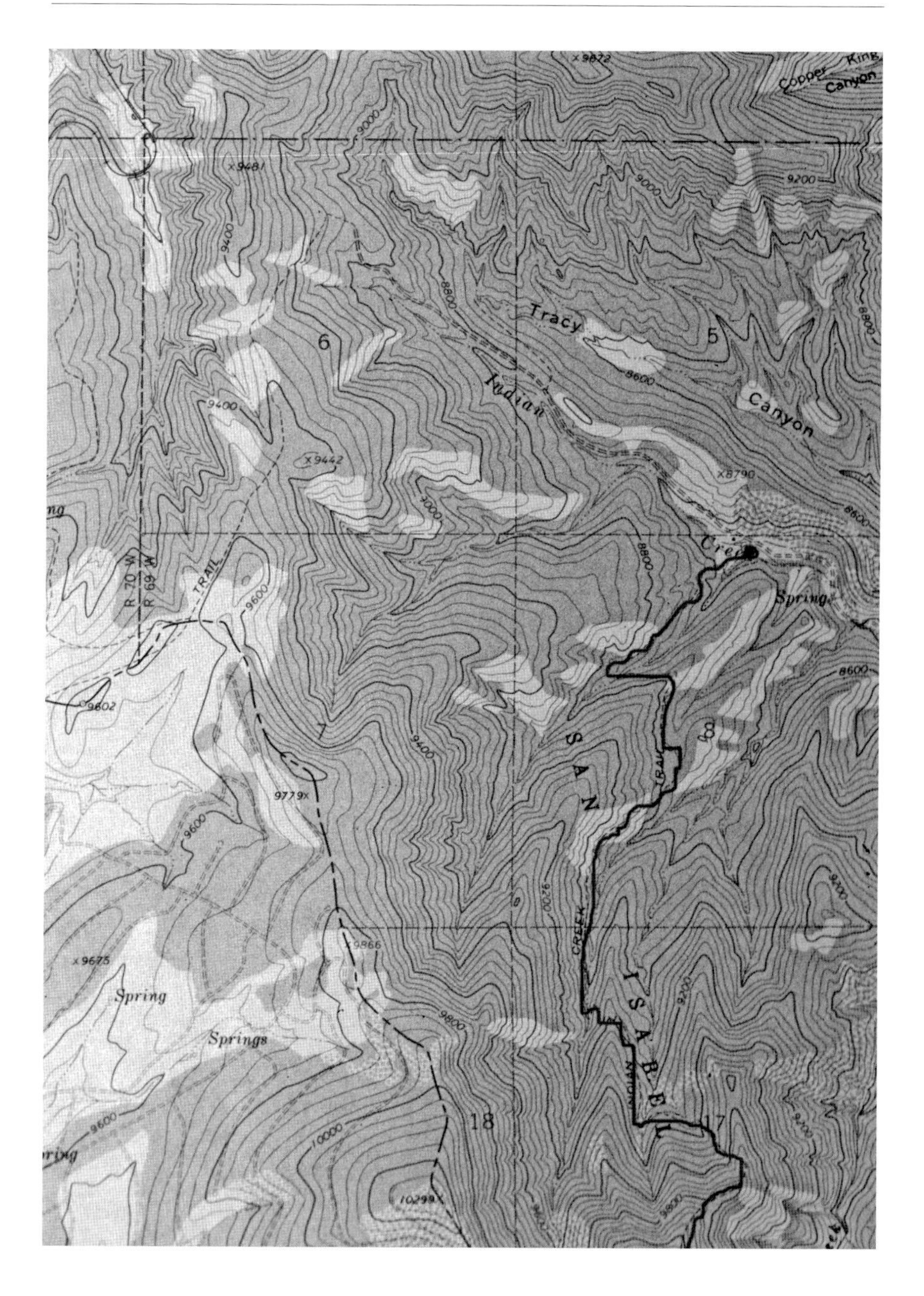

Map 5: Indian Creek Trailhead, north end, beginning at a parking area on Sulphur Springs Road, just beyond the Sulphur Springs Resort. USGS quadrangle: McCarty Park, CO 1982 1:24000.

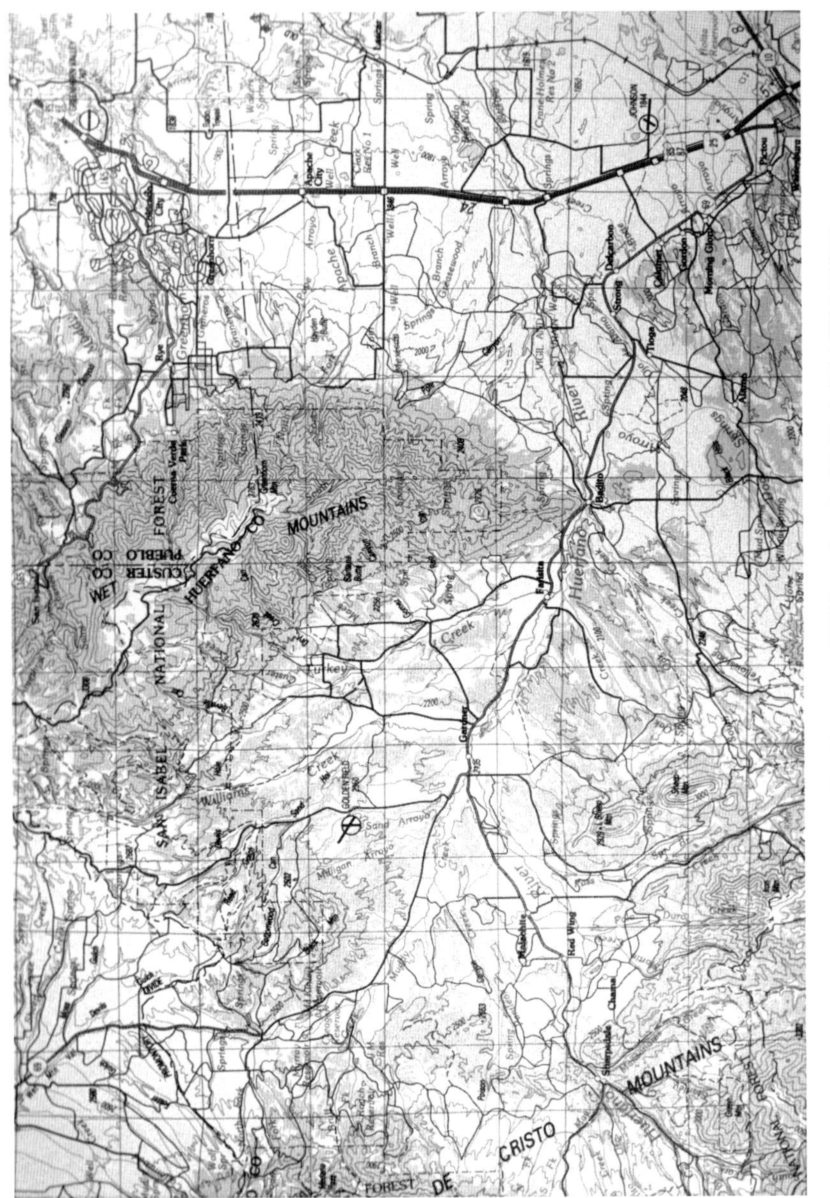

Map 6: La Veta Huerfano River Valley loop; Wet Mountains. USGS quadrangle: Trinidad, CO 1988 1:250000.

Map 7: Greenhorn Peak. Two trailheads are in the parking lot at the end of Greenhorn Mountain Road. The second, which goes to the east and is fairly level, is the Bartlett Trail, which continues around the south side of the mountain near timberline. USGS quadrangle: San Isabel, CO 1969 1:24000.

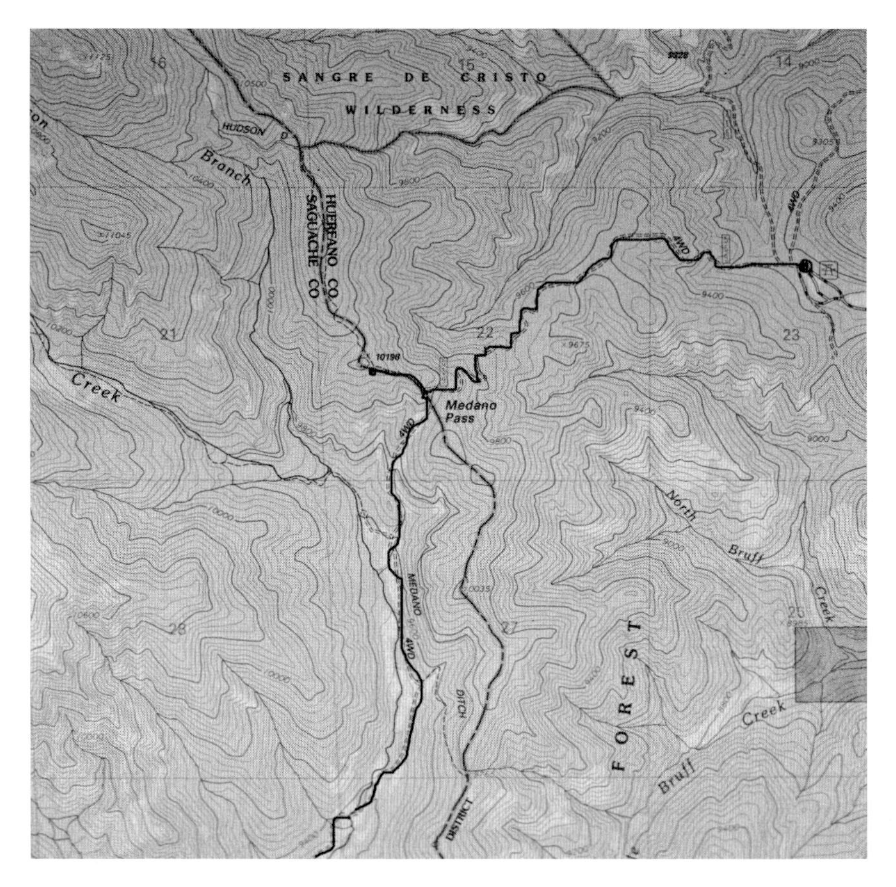

Map 8: Medano Pass 4WD road climbs to the pass, which is not above timberline. On entering the national forest, 4WD is required. At the pass, a spur to the north leads to an opening with good views and a trail on to Mount Herard. USGS quadrangle: Medano Pass, CO 2001 1:24000.

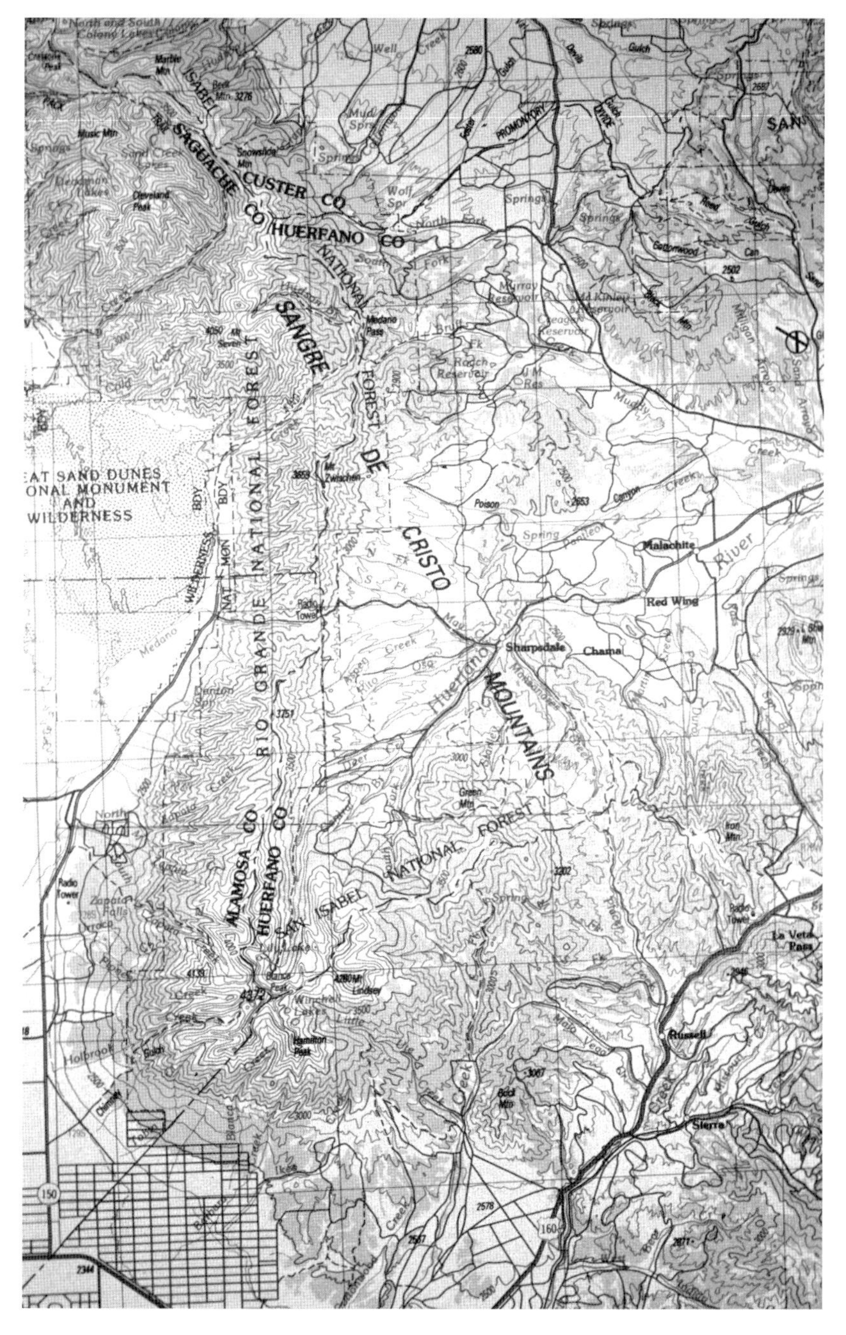

Map 9: Medano and Mosca passes, Blanca Massif and the upper Huerfano Basin, Great Sand Dunes National Park and Preserve. USGS quadrangle: Trinidad, CO 1988 1:250000.

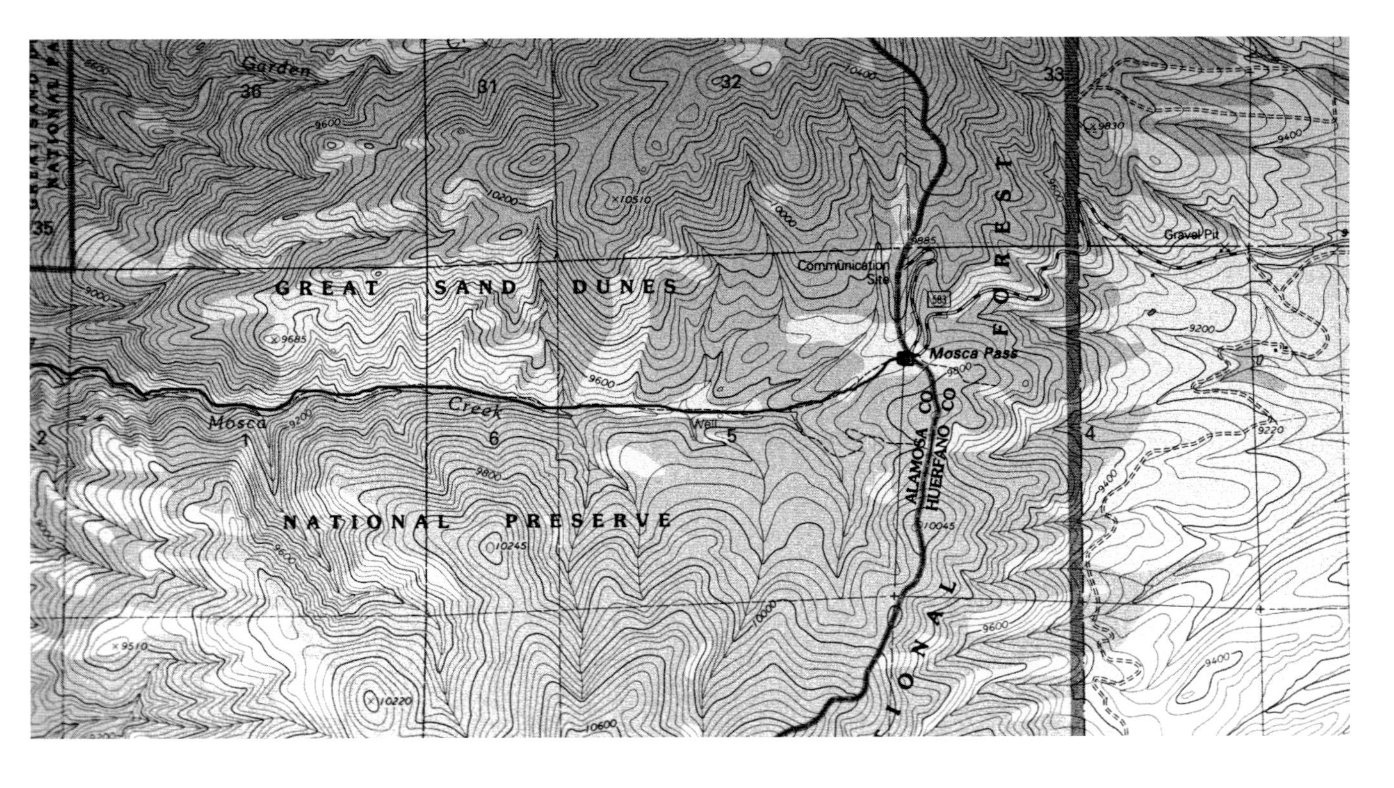

Map 10: A road ascends to the pass from the east. Mosca Pass Trailhead begins at a parking lot on the west side of the pass. The trail along Mosca Creek leads to the sand dunes near the visitor center. USGS quadrangle: Mosca Pass, CO 2001 1:24000.

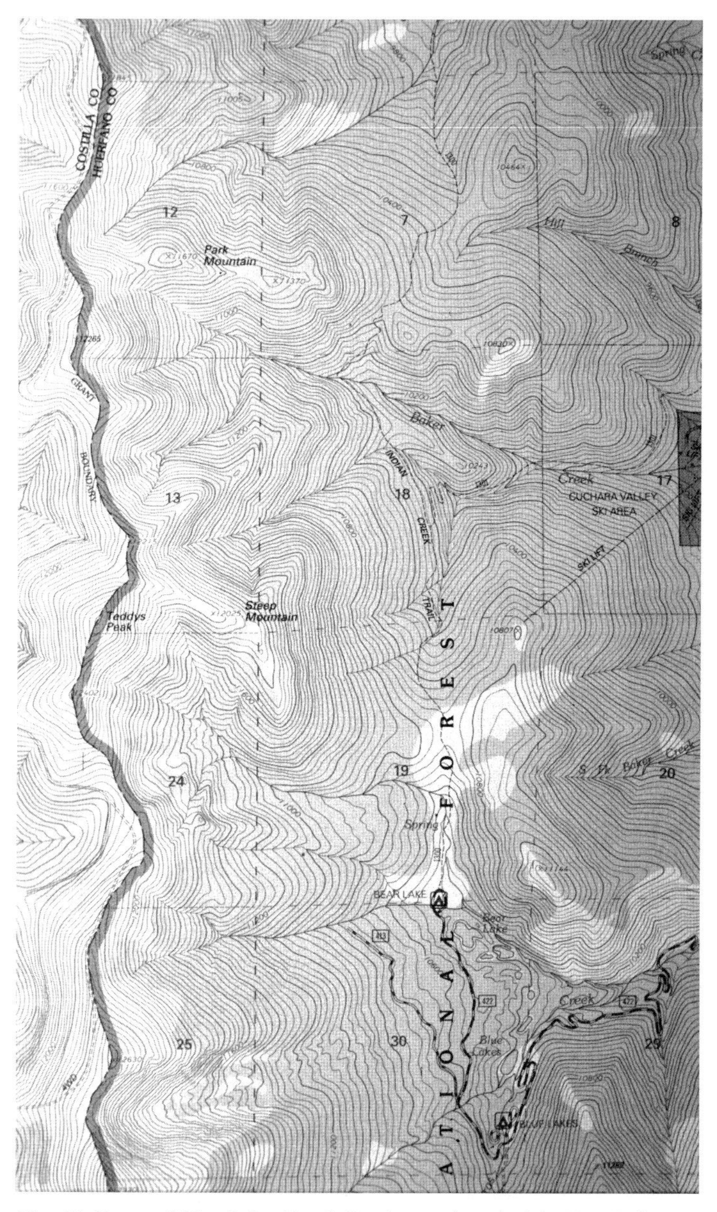

Map 11: Bear and Blue Lakes Road. On the north end of the Bear Lake campground is the south end of the Indian Creek Trail. At the first saddle is the trail to the east to an overlook of the Cuchara Valley. The next right turn is the western terminus of the Baker Creek Trail. USGS quadrangle: McCarty Park, CO 1982 1:24000.

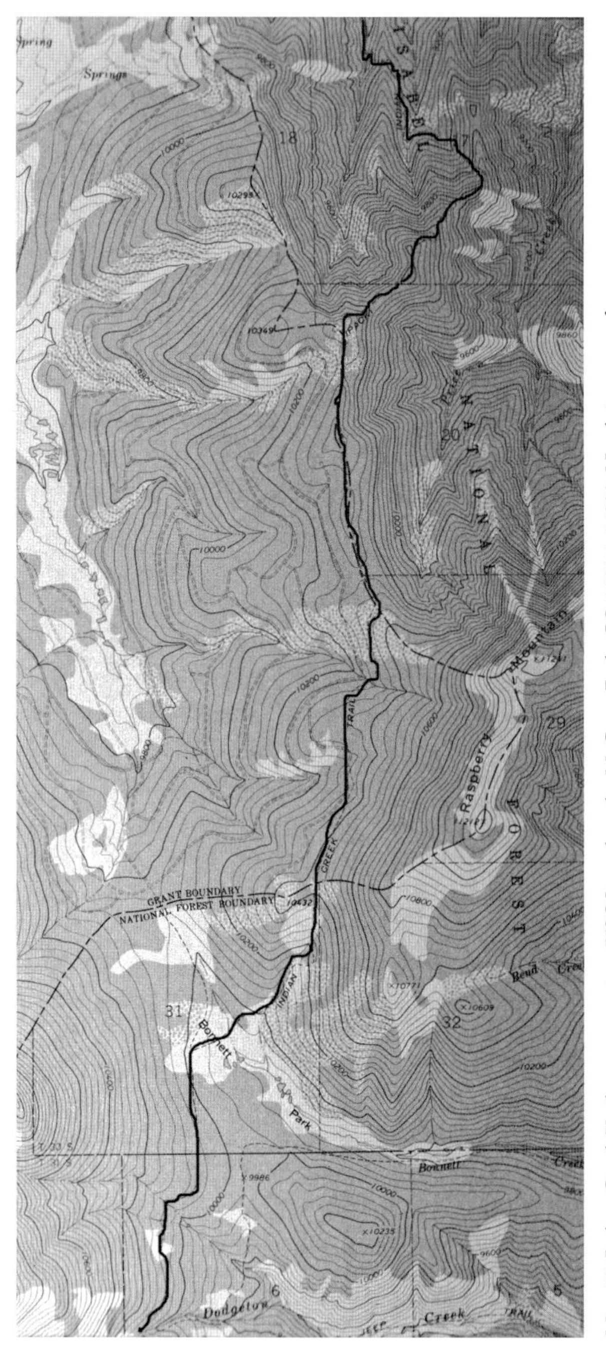

Map 12: Indian Creek Trail, center section. USGS quadrangle: McCarty Park, CO 1982 1:24000. North is at top of page.

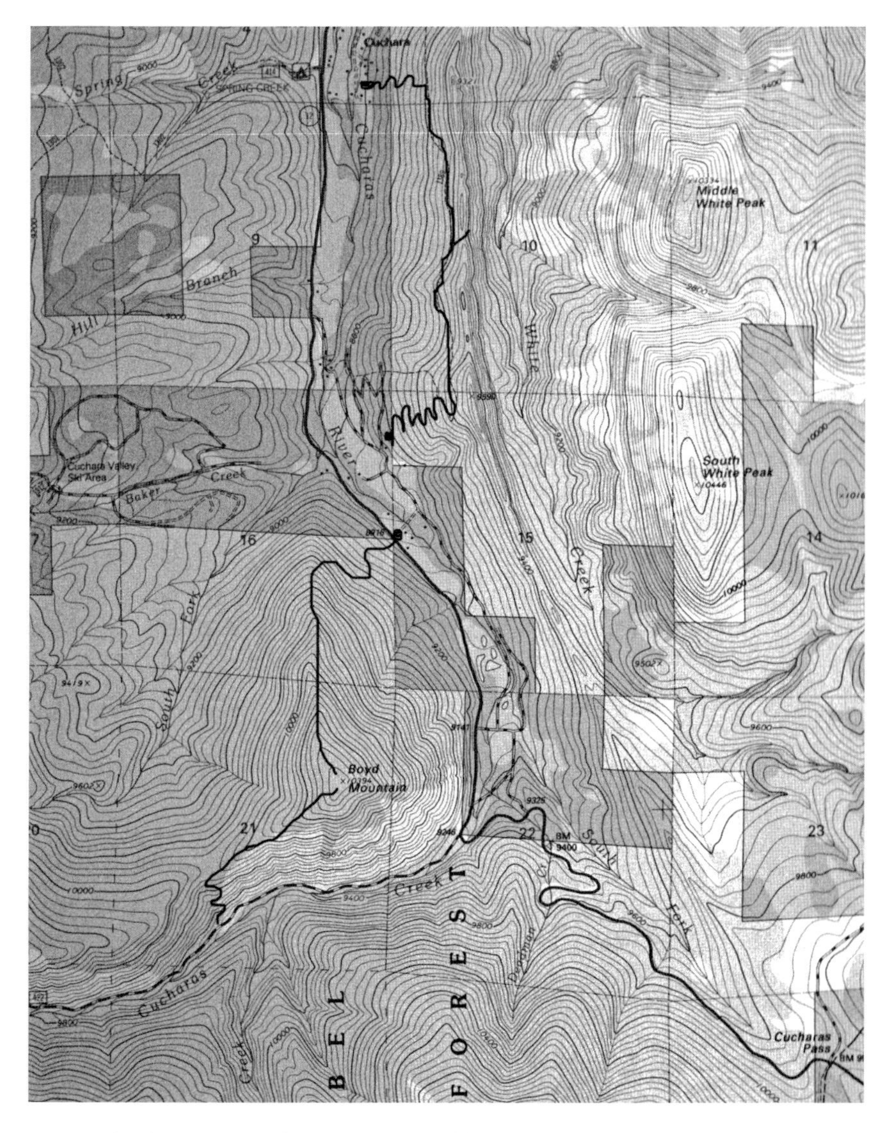

Map 13: On the west side of Route 12 is a feasible cross-country route up the north slope of Boyd Mountain. It can be descended steeply on the south end to Bear Lake Road. East of Route 12, within the Spanish Peaks Subdivision, the Dike Trail begins at a parking lot at the top of Park Road. The trail climbs east, then descends south to Cuchara Village on its north end. Halfway along, a short spur east to the White Peaks Overlook is marked. USGS quadrangle: Cuchara Pass, CO 1994 1:24000.

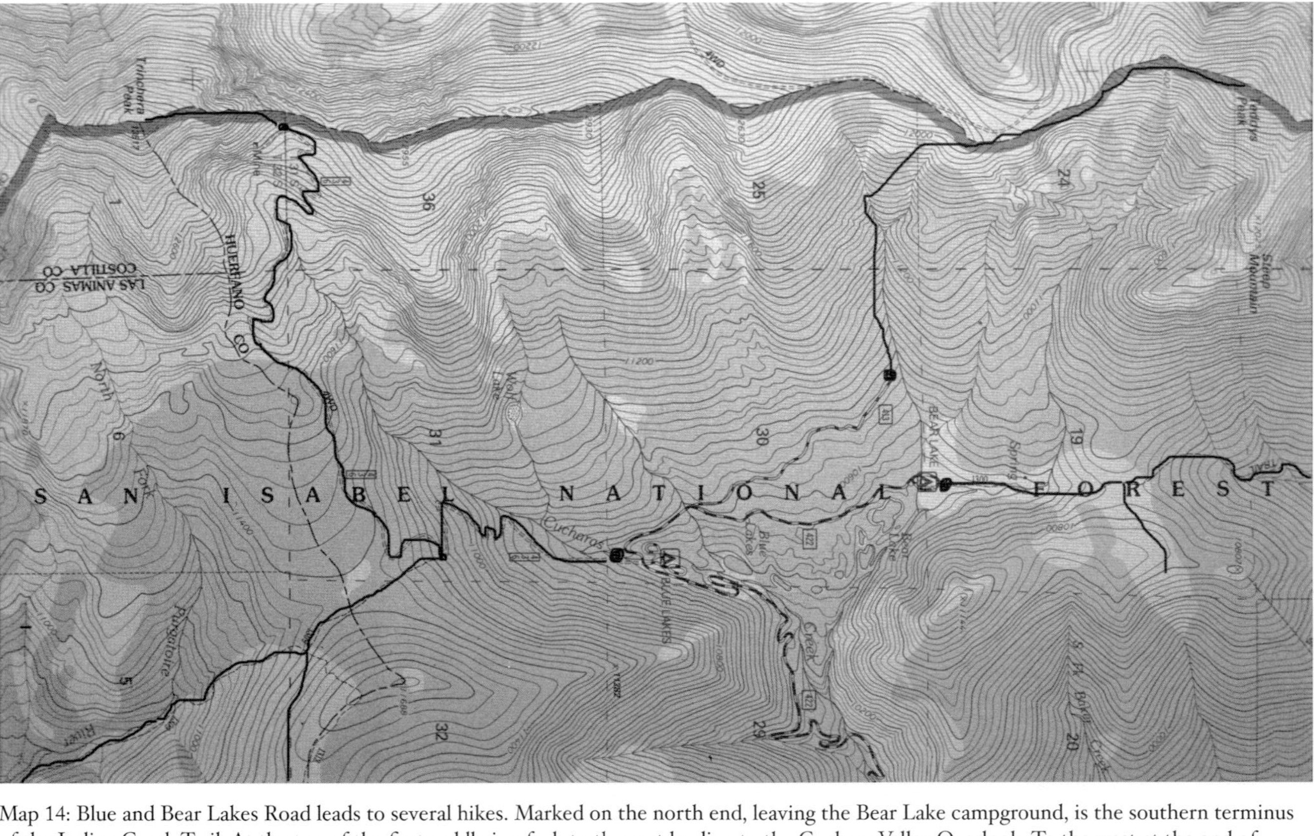

Map 14: Blue and Bear Lakes Road leads to several hikes. Marked on the north end, leaving the Bear Lake campground, is the southern terminus of the Indian Creek Trail. At the top of the first saddle is a fork to the east leading to the Cuchara Valley Overlook. To the west at the end of FR413, one can park and ascend the ridge toward Teddy's Peak. To the south of Blue Lake is the 4WD road/trail to the Trinchera saddle, as well as a scramble to Trinchera Peak. At a marked fork within the timbered slope is a fork to the east that leads to the north fork of the Purgatoire campground. Another branch, just south of the saddle to the east, follows a treeless ridge with a commanding view east and to the Sangre de Cristo Range on the west. USGS quadrangle: Trinchera Peak, CO 1994 1:24000.

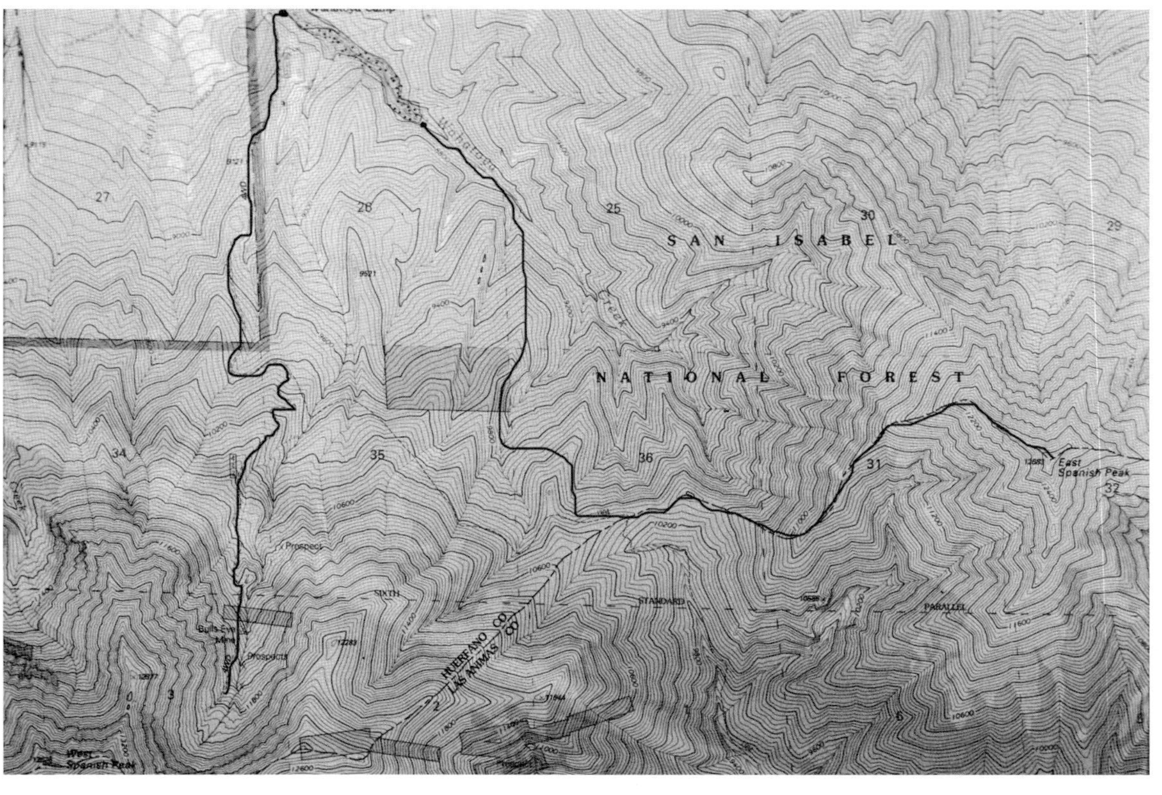

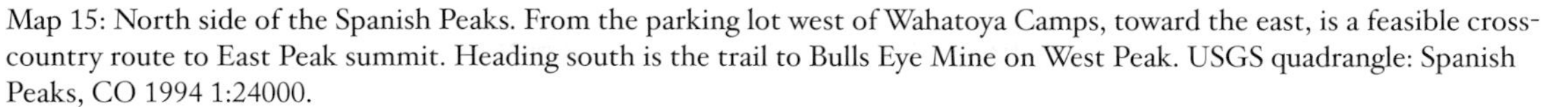

Map 15: North side of the Spanish Peaks. From the parking lot west of Wahatoya Camps, toward the east, is a feasible cross-country route to East Peak summit. Heading south is the trail to Bulls Eye Mine on West Peak. USGS quadrangle: Spanish Peaks, CO 1994 1:24000.

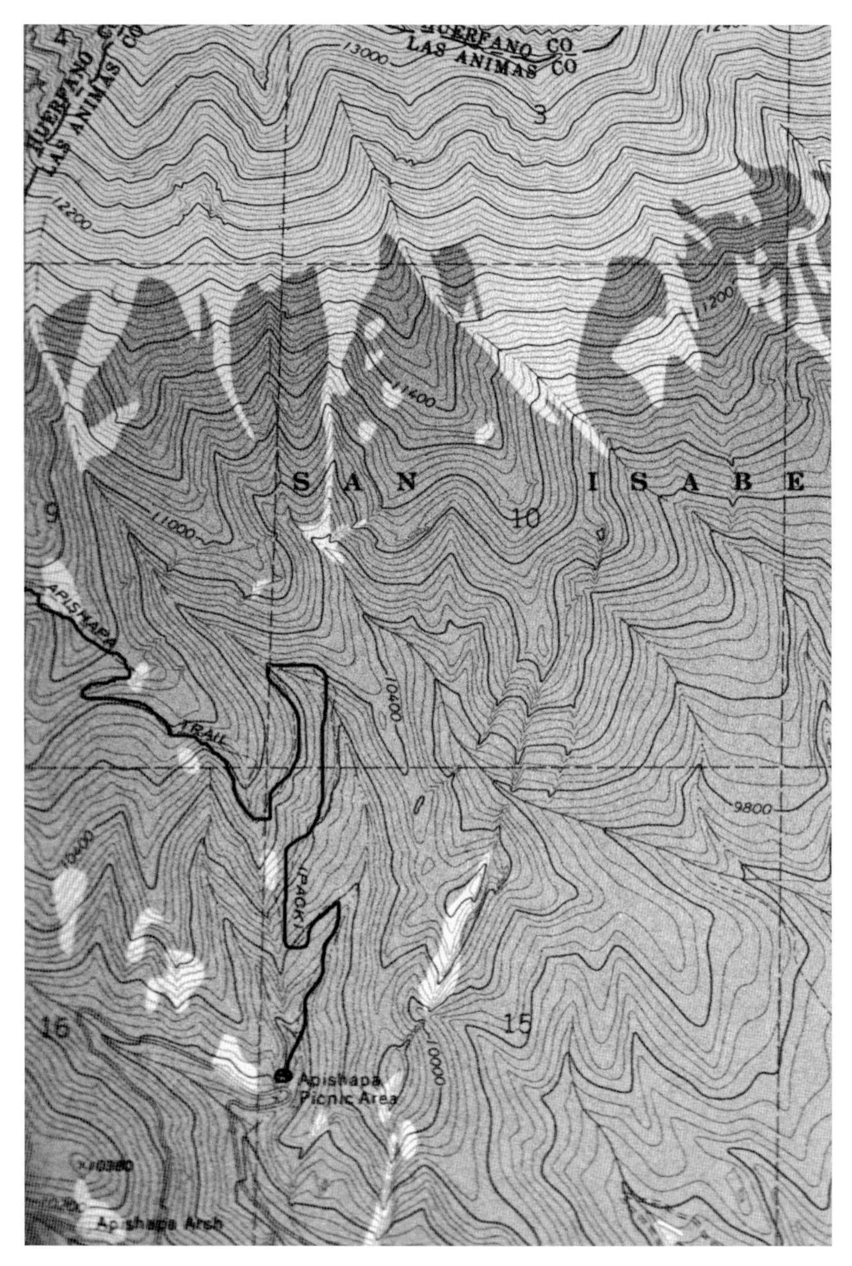

Map 16: Apishapa Trailhead begins to the north of the parking area and around a bend in the road to the north. A short upward trail from the parking area leads to the trail entrance across the road. The trail connects at the west end with the West Peak Trail that leaves from Cordova Pass. USGS quadrangle: Herlick Canyon, CO 1971 1:24000.

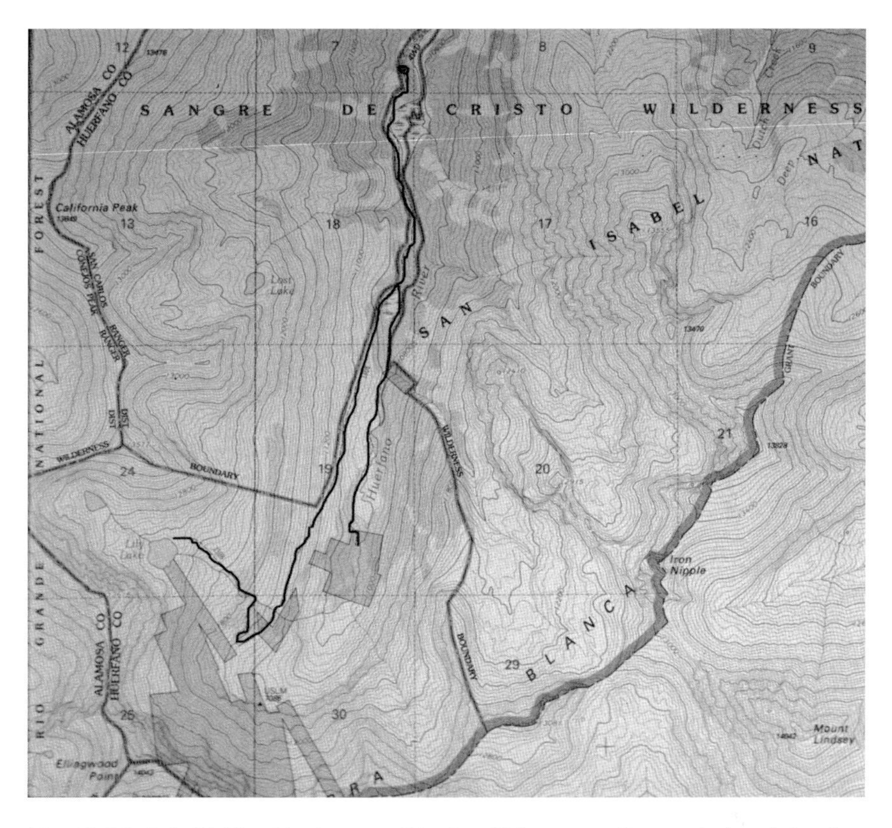

Map 17: Lily Lake Trailhead in the upper Huerfano Valley, proceeding south. At a fork, take the right or southwest-bearing fork that ascends the flank of the range and ends at Lily Lake. Bearing straight or southeast, the level trail leads to an old logging mill site. USGS quadrangle: Blanca Peak, CO 2001 1:24000.

CHAPTER 7

SCENIC DRIVES

SOURCES

Exploring the roads of this varied region is greatly facilitated by owning a Colorado State road atlas. Several are published and updated each year. Most commonly available is DeLorme's *Colorado Atlas and Gazetteer*, which divides the state into about 100 pages of maps with contour lines and includes nearly all roads and trails. *Pierson Guides Colorado: Recreational Road Atlas* is good but less available. U.S. Geological Survey (USGS) quadrangles can be purchased at the USGS office in Denver, at outdoor stores, and at the Visitor Center at Great Sand Dunes National Park in the San Luis Valley. Specific maps are available for the Sangre de Cristo range as well. County maps are available for free at some places of business, museums, and county offices in La Veta, Walsenburg, and Trinidad. The DeLorme Atlas is the source for rural road numbers used below.

Note that museum days and hours of operation are variable and often seasonal. Ask at visitor centers or chambers of commerce offices for current information.

ROAD CONDITIONS

Road conditions range from paved highways, to gravel roads easily navigated by ordinary cars, to rougher roads requiring higher ground clearance, to routes where four-wheel drive (4WD) is mandatory. They are not specifically ranked below, because unpaved road conditions vary year by year. Rural road maintenance may have been either neglected or upgraded since my last visit. I will try to alert you to roads requiring high ground clearance or 4WD. Be aware of your own limitations, as well as those of your vehicle. On primitive roads, remember the adage "four wheel drive means getting stuck in more inaccessible places." *A final note on weather*: When traveling dirt or sandy roads on up-down grades, be mindful of the weather. Some muddy roads become nearly impassable.

TRESPASSING

Some landowners are quite touchy about public trespass and threaten prosecution. Pay attention to postings and respect private property. See Appendix 2 for notes on Colorado trespass regulations.

ROADKILL

Over the past three decades, roadkill has overtaken hunting as the number one human-induced cause of direct death to wild animals on land. According to the Humane Society of the United States, between 500,000 and one million vertebrate animals perish on our roads each day. Make an effort to avoid adding to this carnage. Small animals are ill equipped to judge automotive speeds, and remember that large animals exact posthumous retribution on your vehicle. Many animals are especially active at dawn and dusk.

DESCANSOS

While driving this countryside, notice occasional *descansos* at the roadside. These are little memorials frequently placed at the site of a highway death, but not necessarily for that reason. They are especially common in the Southwest but are appearing now in many states. Descanso means "resting place" in Hispanic folk culture. Before the automobile, village funeral processions meant carrying a heavy handmade casket some distance to the cemetery. Where the pallbearers rested, a small cross was placed to mark the "interrupted journey," which also symbolizes life and death.

On remote roads in the Sangre de Cristos, perhaps at the site of a fine view, you can find these little shrines. When it can be done safely, I stop to look. Many are elaborate examples of folk art complete with a handmade wooden or metal cross, framed poetry, photos, prayers, and the names of family members and objects they contributed. Even the simple ones show the same love. Somehow, when you stand in front of one you can feel the spirit of the families and their connection with the landscape.

INDEX OF DRIVES

The drives are organized into four regions. Each begins with a loop drive or major destination. Drives that branch off regional loops are lettered as below.

South:

1. Highway of Legends loop, Route 12 and I-25

1A. Cordova Pass Road to Aguilar

1B. Jarosa Canyon, 42.0 Road and Wet Canyon 31.9 Road

1C. Walsenburg to La Veta, the back way

1D. Sulphur Springs Road

Central:

2. La Veta to Huerfano Valley Loop: Yellowstone Road to Route 69, Badito, Gardner, to Pass Creek Road

2A. South Oak Creek Canyon

North:

3. Greenhorn (Wet Mountains) Loop; Mission Wolf Sanctuary (north of Gardner), Ophir Creek Road (upper Huerfano Valley off 69 to local 634 to Greenhorn Peak)

West:

4A. Medano Pass Road to Great Sand Dunes National Park and Preserve

4B. Mosca Pass, with trailhead for Great Sand Dunes National Park

4C. Upper Huerfano Basin, 4WD or walk; Lily Lake Trailhead; Huerfano State Wildlife Area (SWA)

4D. South (Old) La Veta Pass Road

SOUTH

DRIVE #1: HIGHWAY OF LEGENDS LOOP, ROUTE 12 AND I-25 (SEE MAPS 1 AND 2)

Features: This Colorado Scenic Byway and National Forest Byway is recommended as a general get-acquainted tourist drive. While it is entirely paved, do not plan to make fast time on the route, as its curves and grades enforce a stately pace. On its north end, Route 12 arises at Colorado Route 160, 5 miles northeast of La Veta (ca. 7000-foot elevation), makes its way over Cuchara Pass (ca. 9990-foot elevation), then descends to Trinidad (6200-foot elevation). The drive traverses spectacular scenery and historical towns. A chamber of commerce booklet, *Scenic Highway of Legends*, is available free or for a small fee and describes the towns, landmarks, and legends along the drive in some detail. Obtain a copy from the chamber of commerce office in Walsenburg, La Veta, or Trinidad before your trip. The 116-mile loop can be driven in a half day, but be sure to allow a whole day for stops and sightseeing.

Approach: From the north: Leaving Walsenburg on Route 160 west, follow signs to La Veta, a left turn, which is the beginning of Route 12 (milepost 0.3). La Veta is a post–Civil War town that, for its small size, features several cultural amenities. The pleasant city park includes the Spanish Peaks Arts Council gallery. Art shows are displayed throughout the year. A block to the south is the Francisco Fort Museum, a complex of several buildings that includes a log schoolhouse, Colonel Francisco's adobe plaza fort building, a historic saloon, a post office, and a mining room. The museum features a rich collection of pioneer and American Indian life artifacts; tours are available.

Follow Route 12 southward over Cuchara Pass, into Las Animas County, and continue descending to Trinidad. Walsenburg to Trinidad is a semicircular drive, connecting back to Walsenburg straight north via I-25.

Mile 0.0: Beginning at La Veta, on Main Street, follow Route 12 signs that zig and zag out of town, traveling south. Immediately looming to your left are West and East Spanish peaks. The southern Sangre de Cristos, also called the Culebra Range, are on your right. Leaving La Veta you will pass Grandote Peaks Golf Course, named after a legendary American Indian who came to the area to prospect for gold. On your right will be an isolated volcanic feature called Goemmer's Butte.

Mile 4.7: You will pass spectacular dikes radiating from West Spanish Peak. Profile Rock has a craggy ridge with profiles variously alleged to resemble George Washington, Thomas Jefferson, Martha Washington, an American Indian, or a small train.

Mile 6.7: The giant angles of Devil's Stairsteps are easily recognized after you have passed it. Near each of these formations there is space to pull off the road to photograph (see Chapter 4, Geology, for a description of these features and the Spanish Peaks).

Mile 9: With a sharp right bend, you pass through a gap in the Dakota Wall. Although it resembles a dike, the wall is an entirely different feature. First, it does not radiate from the Spanish Peaks, or any mountain, but instead runs north-south, tangential to them. Second, it is made of a resistant bed of sandstone, a very different material from the igneous dikes. This very hard formation was upturned during the formation of the Culebra Range to the west (Fig. 39). You will see the Dakota Wall again on the other side of Cuchara Pass

Fig. 39: Route 12 looking south toward Cuchara. "The Gap" in the Dakota Wall appears in the middle distance.

as you descend into Las Animas County. Geologists have identified the Dakota Wall outcropping occasionally over hundreds of miles running north-south and aligned vaguely with the Front Range.

In the Cuchara Valley, air movements are constricted at the gap. Local weather or cloud development often changes character abruptly here. Continuing your climb, the valley gets narrower and the road sinuous.

Mile 11.8: The village of Cuchara (8600-foot elevation) appears on the left. Its main street, locally called "downtown Cuchara," extends about a half block and offers lodging, restaurants, ice cream shops, and gift shops. The theme is "Old West."

The road climbs steadily. About 1 mile past Cuchara, at a left pull-off, a red bed road cut appears on the right that exposes a small gray dike in cross section. It was a high-temperature intrusion, and the heat-altered rock can be inspected at the boundary of the surrounding red rock (see Chapter 4, Geology).

The Spanish Peaks Subdivision heaves into view on the left, and then the Cuchara Mountain Resort and subdivision on the right. The resort and ski area have an interesting history. Originally built in about 1982 on a permit from the U.S. Forest Service, the ski area has been plagued by a general lack of financial viability. Under various ownerships, it operated on and off for 14 years and ceased operation in 2000. Despite the fact that it was the only ski area accessible from the Front Range without going over a pass, it never became competitive. Then, after a decade of neglect a new initiative began in the summer of 2010 when a buyer announced a plan to reestablish a resort operation. This time around it will include year-round attractions, conference facilities, and a plan to turn over management to a nonprofit board comprised of local citizens. While my initial draft stated, "It seems unlikely that these lifts will ever return again," I wish this new initiative success.

You skirt Boyd Mountain on the right. In the fall, the Trinchera elk herd may be found on its flanks.

Mile 15.2: Where the highway bears right, and then makes a sharp left bend, a forest road to Cuchara River Recreation Area enters on the right. Although occasionally freshly graded, the usually washboardy, 5-mile gravel side road climbs along the Cuchara River and will bring you to campgrounds and trails near Blue and Bear lakes. The lakes are popular trout fishing destinations. This is a San Isabel National Forest Recreation Area, and day use and camping fees are collected. Colorado fishing licenses are available in Cuchara, La Veta, and Walsenburg. From Blue Lake, a primitive 4WD road proceeds southwest in the direction of Trinchera Peak.

Route 12 continues to climb more steeply with a few switchbacks.

Mile 17.6: You reach Cuchara Pass at 9994 feet. Stop there to see more open vistas both north and south. At the pass itself, a road forks to the left (east) and continues up to Cordova Pass and on down to Aguilar near I-25 (see Drive 1A for a description). A half-mile side trip up this road brings you to the Farley Wildflower Trail and parking area. There are signs and an interpretive garden that illustrates local flowering plants (Fig. 40). Looking north, there are great views over the Cuchara Valley.

Following Route 12 from Cuchara Pass, you descend south into Las Animas County. The landscape is more open, arid, and less forested. Considerable gently rolling and picturesque open range allows a panoramic view of northern New Mexico. You are now in the watershed of a river known as the Purgatoire, or Purgatory, also referred to locally as "Picketwire." Out on the plains, the Purgatoire connects with the Huerfano and Cuchara rivers, all eventual tributaries of the Arkansas to the east of Pueblo.

Mile 24.7: The road curves sharply right, and North Lake appears on the left. It is a state wildlife area boasting good trout fishing. Farther on, **Mile 25.3** is Monument Lake, named for a pointed boulder of Dakota Wall sandstone that once emerged from its water. Sometime in 2001, the "monument" toppled into the lake, reminding us of the forces of erosion and gravity. Camping facili-

Fig. 40 (left): Wildflowers at the Farley Wildflower Overlook east of Cuchara Pass. Boyd Mountain is seen in the middle ground, beneath the Sangre de Cristo crest on the horizon.

Fig. 41 (below): Prairie coneflowers (*Ratibida pinnata* (Vent.) Barnhart) on the road to Trinidad near North and Monument lakes.

ties, cabins, a boat launch, and a restaurant are available at Monument Lake recreation area (Fig. 41).

Mile 30.1: Near Stonewall you pass through the Dakota Wall again, and it can be seen in several places as you descend this road. Route 12's direction is now mostly due east.

Mile 33.9: The New Elk mine appears on the right. One of the newer mines in the vicinity, it was most recently a coal processing facility.

Next you pass through a series of original Spanish plaza towns with a 140-year history of pioneer agricultural activities as well as coal mining. Vigil was named for Juan Vigil, a county sheriff in the 1860s.

Mile 34.9: Off to the left (north), one can see the "House on the Bridge." The house was built there after the bridge and its road alignment were abandoned. You will encounter Weston, Cordova Plaza, Segundo, and rows of hundreds of picturesque coke ovens at Cokedale, a national historical site.

From various vantage points on the highway, one can see ahead (east) the flat-topped Fisher's Peak that forms a massive and handsome backdrop to the city of Trinidad. The flat peak is capped by a bed of lava, a remnant of a once-widespread regional deposit. Unfortunately, there is no longer any public access to its commanding summit. The peak, once called "Raton Peak," was renamed for Captain Waldemar Fischer, who climbed it for orientation. Later, he led a cavalry during the New Mexican conflict of 1846.

Mile 59.6: Before arriving at Trinidad, you descend a long grade and Trinidad Lake State Park will be on the right (south). The dam impounding the Purgatoire River is owned by the U.S. Army Corps of Engineers (USACE) (Fig. 42). The state leases the surrounding land and, together with some state land, manages recreational programs including camping, picnicking, fishing, hunting, and boating. Fees are collected for park entrance and for camping. The upper entrance leads to a small visitor center (within 0.5 mile) that has introductory demonstrations of ecology and human history and information about the building of the impoundment. Some publications and brochures are available, and several trails ranging in length from 1 to about 4 miles can be enjoyed. From the visitor center, one drives east along a ridge to a picnic area in mature pinyon-juniper woodland. Each site has nice views north and east toward Fisher's Peak, over the reservoir basin (Fig. 43). At the end of the loop is access to the well-equipped campground.

Two rock circles within the turnaround loop are displayed as prehistoric archeological remains of an American Indian village. However, there have been no test trenches or published studies that would confirm their authenticity.

The imposing Trinidad Dam, built by the USACE, once impounded a sub-

Fig. 42: Trinidad Lake during a dry year, with Fisher's Peak on the horizon.

Fig. 43: Trinidad Lake with Fisher's Peak on the right, as seen from the picnic loop road in Trinidad Lake State Park. More water is present, but the lake level is below the basin design capacity.

stantial reservoir that can be reached by an entrance road from Route 12, about 1 mile east of the entrance to the visitor center. The road across the 6860-foot-long dam affords fine views of Fisher's Peak to the east and the remaining water to the west. All that remains of the impoundment now is a small lake at the bottom of the basin.

To build the lake in the 1970s, railroad tracks were moved, 7 miles of Route 12 and six towns were relocated, and a number of Apache archeological sites were drowned. For a few years, the lake was at capacity as a water storage structure, holding 125,967 acre-feet of water, allocated among flood control, irrigation, sediment control, and recreation. At capacity, its surface area is at least 900 acres (figures for surface areas and acre-feet vary among published sources). Presently, the 200-foot-high dam, piled from 8 million cubic yards of earth and rock, sits high and dry like the Great Wall of China. Although the water level fluctuates depending on wet and dry years, the main issues involve water rights.

In 1948, Colorado and Kansas divided water rights according to the Arkansas River Compact. The Purgatoire River is a major tributary of the Arkansas River that, in turn, flows through eastern Colorado and Kansas on its way to the Mississippi. Unforeseen in those negotiations was the subsequent drilling of about 2000 wells in Colorado's Arkansas Basin, as well as the building and management of the Trinidad Reservoir. For more than 25 years, Kansas challenged Colorado's alleged overuse of its "drawing rights," which led to ongoing monitoring of water allocation through the offices of the "Special Master," a water-specialist lawyer whose findings were reported to the U.S. Supreme Court. By the late 1990s, the court, in a mixed decision, had recommended that Colorado be assessed financial penalties for overdrawing its agreed allocations. Litigation is ongoing at this time.

In the end, Trinidad Dam's principal remaining function is flood control. The Purgatoire has produced six catastrophic floods through Trinidad between 1883 and 1965, but the townspeople can now rest assured that floods are not part of their future.

Mile 62.8: After zigzagging through the western outskirts of Trinidad, Route 12 passes under the I-25 overpass. Just beyond the overpass, stop at the visitor center on the left. Knowledgeable staff can organize your visit to the region. Note that some of the attractions are not open throughout the year. To park, you may have to circle the block once because the entrance is not intuitive. Do not attempt to enter the parking area with a large rig.

In addition to the usual amenities, the town culture exemplifies the variety that reflects the starkly contrasting western landscape itself. Trinidad's respected junior college has one of the largest gunsmithing programs in the country. In other circles, the town is noted as the "sex-change capital of the world." For decades, the late Dr. Stanley Biber, who died in January 2006, had a general surgical practice for which he gained local renown. He also performed about 5000 gender reassignments, more than any other clinic in the country.

Trinidad's pride in its striking architectural heritage can be seen on Main Street. In a compact three- to four-block area are the Trinidad History Museum buildings, where one can tour the 1882 Bloom Mansion, the Baca House, and the Santa Fe Trail Museum. Down the block is the A.R. Mitchell Memorial Museum of Western Art, featuring an exceptional western art collection and traveling exhibits. Each year, Trinidad celebrates its heritage as a major stop on the Santa Fe Trail. Several interesting restaurants and shops occupy the same neighborhood. Nearby on Commercial Street is the Old Firehouse No. 1 Children's Museum, of interest to the whole family.

On the west side of town, on the Trinidad Junior College campus, is the Louden-Henritze Archeology Museum. Geological and archeological history as interpreted through fossils and human artifacts is very well presented.

For an exceptional view of the town and the region, ask for directions to Simpson's Rest, a short drive up to a commanding bluff top on the northwest edge of town (Fig. 44). No signs identify the route. At an elevation of 6478 feet, the overlook displays fine views in all directions that include Raton Pass to the south, the city, Fisher's Peak, and mountains to the west. As an Army scout, Simpson loved this promontory on which he once saved himself from Indians in hot pursuit.

If you are traveling the "Highway of Legends" as a loop road, take I-25 north toward Walsenburg.

Mile 75: The Ludlow Massacre Memorial site is just off I-25, about 12 miles north of Trinidad (23 miles south of Walsenburg) at a marked exit. Head straight west about 1 mile toward a grove of trees, a fence, a covered pavilion, and a parking area. This small site is very significant in the history of American organized labor.

At the site's center stands a large granite memorial. On its east-facing side are the names of 11 women and children who were burned to death in a pit beneath the tent in which they were hiding. Facing west are granite carvings of a coal miner and his wife. Nearby one can descend into a concrete bunker where the killings took place.

Unorganized miners in those days lived in ethnically segregated tent villages and were subject to economic and social repression and terrible working conditions. Between 1902 and 1917 about 400 miners were killed in the region by

Fig. 44: The city of Trinidad, with Fisher's Peak on the horizon, as seen looking toward the southeast from Simpson's Rest.

coal dust explosions. In 1913, from among several mining companies operating in the area, the United Mine Workers of America decided to strike against the anti-union coal company, Colorado Fuel and Iron. The Colorado State militia was called out on behalf of the company to restore order. The resulting armed conflict ended with the militia and strikebreakers destroying and looting the camp. In later years, nationwide public outrage led to reforms that were a turning point in the history of American labor relations.

Beneath the pavilion roof are explanatory signs detailing the history of the Coalfield War of 1913 to 1914. This heroic chapter in the history of American labor is too complex to detail here. Senator George McGovern's doctoral thesis, *The Great Coalfield War*, remains the definitive analysis and history of this event and the wider conflict (see McGovern and Guttridge, 1996). Historically minded visitors will find their visit to be time well spent. Families with union connections may find it especially moving.

Mile 100: Arrive at Walsenburg. Take the first (south) of three exits from I-25, which is designated as routes 85/87. This becomes Main Street and then Walsen Avenue in the northern half of town. Just east of Main, on Route 10 or 5th Street, is the Walsenburg Chamber of Commerce. Housed in the picturesque old railroad depot, you can find much information on the town and Huerfano County. On the west side of Main, on 5th Street, is the attractive stone courthouse building. Just behind it (to the west) is the Walsenburg Mining Museum, housed in the century-old former jail.

Mile 101.5: To continue the loop drive, at the stoplight at Main and 7th streets, follow Route 160 west.

Mile 104: Lathrop State Park entrance is on the right (north). As Colorado's first state park, it dates from 1962. This fee area features a visitor center with natural history exhibits, two prominent fishing lakes, some ponds, camping facilities, and nature trails with scenic views of the valley and the Spanish Peaks.

Mile 112: Route 12 toward La Veta and Trinidad originates as a left (south) turn.

Mile 116: Arrive at La Veta.

DRIVE #1A: CUCHARA PASS TO AGUILAR, VIA CORDOVA (APISHAPA) PASS (SEE MAPS 2 TO 4)

Features: The road runs east-west, providing views across the south side of the Spanish Peaks and out to the Great Plains in several places. The road is gravel and generally well graded. If driving a low-ground-clearance car, be vigilant for larger rocks, especially when driving down the east side of the pass. A great transect of the elevational changes in vegetation communities is exposed for your appreciation.

Approach: You can begin this road from the east or west. Either way, it is easy to find. From Cuchara, drive south up Route 12 to Cuchara Pass.

Mile 0: At Cuchara Pass summit, turn left (east) onto FR 415 and drive on a gravel but usually well-graded road. At the first half mile, stop at the Farley Wildflower Overlook on your left. The surrounding meadow has a nice display of wildflowers in the summer and fine views to the north. For the next 20 minutes, the road climbs steadily through meadows lined by aspens, then spruces and firs on an elevation gradient where tree species replace each other according to their elevational tolerances.

Mile 6: Arriving at Cordova Pass (elevation 11,264 feet, 37.34833°N, 105.02500°W), you will find a broad, well-maintained parking place at the top. This is a fee area for anyone camping or using the trails (see Chapter 8, Trails, for trails leading toward West Peak). The most popular trail to West Spanish Peak begins here and runs north. You will find a good view to the west over the Sangre de Cristo Range.

Driving east, the designation becomes 46.0 Road. It begins as a broad gravel road with a steep descent that becomes narrow in spots. In a few places there are pullouts, some of which have good views. You traverse a landscape that overlooks the western plains and flat-topped mesas to the east and south. Numerous dikes are in view in different angles, and the road passes through a tunnel in one dike (Fig. 45).

Mile 14.7: Reach the Apishapa Trailhead on the left. The parking area is around the next bend on the right. It is a wilderness trail that goes to the saddle between the Spanish Peaks. The road proceeds downward through dense growth of spruces and firs. Soon the road reaches the bottom of the Spanish Peaks uplift. It gradually and gently descends along the Apishapa River into a broadening valley.

Upon reaching the base, the road is fairly level and the landscape opens up to a Great Plains habitat. Four miles to the west of Aguilar you reach pavement.

Mile 38: Arrive at the small town of Aguilar. Within 1 mile, join I-25 and travel south to Trinidad (16 miles) or north to Walsenburg. Driving north, you reach Walsenburg at mile 58. Arrive at La Veta at mile 76.9. Arrive at Cuchara at ca. mile 90.

Fig. 45: A tunnel through the dike on the road to Aguilar.

DRIVE #1B: JAROSA CANYON AND WET CANYON (SEE MAPS 2 AND 4)

Features: This drive enters a landscape of rolling scenery on the south side of the Spanish Peaks hinterland north of Route 12 in Las Animas County. Jarosa Canyon–Wet Canyon Road gradually climbs through a valley with pinyon-juniper and scrub oak, then rises through a young ponderosa pine forest. On the Jarosa Canyon side, the Spanish Peaks can be glimpsed in several places. The vegetation sequence reverses on the way south down to Weston. The area is sparsely populated until one approaches Weston.

The climate on south-facing slopes is drier than that of predominantly north-facing slopes and the vegetation is less luxurious and diverse. Ranching and coal bed methane extraction are present. The latter distracts little from the attractive countryside.

Approach: See the description for Drive 1A. The road (42.0 Road) begins as a fork off the road between Aguilar and Cordova Pass. It can be reached from Aguilar, near I-25, or from Weston on Route 12. It is described here as a circuit beginning at Cuchara Pass via Cordova (Apishapa) Pass, to Weston, west to Stonewall, thence back to Cuchara Pass.

Mile 0: Cordova Pass.

Mile 18: Reach Jarosa Canyon road (42.0 Road), entering on the right at an oblique angle. (From Aguilar, this fork is 12.3 miles west.) It climbs at a modest rate for about 11 miles traveling west.

Mile 35: At the high point (8661-foot elevation), once called "Wet Canyon Pass," the road turns south and is renamed Wet Canyon Road. Descend gradually east, then south again.

Mile 50.8: Arrive at the town of Weston on Route 12 opposite "Weston Supply/General Merchandise/U.S. Post Office."

As a return loop, take Route 12 to the right (west) and proceed through Stonewall, returning to Cuchara Pass. A complete 85-mile circuit is possible, with a 40-minute lunch stop, in about 3.5 hours.

Bosque del Oso State Wildlife Area: For an all-day outing to this natural area, a side trip with hikes can be enjoyed in this undeveloped site. At Weston, from the intersection of Wet Canyon Road and Route 12, turn east one block, then south into the marked entrance to the state wildlife area (SWA). You need a $10 permit to enter SWAs, and permits are not available on site. Remember that there are no visitor facilities on SWAs (see Chapter 2, Federal Lands).

DRIVE #1C: WALSENBURG TO LA VETA, "THE BACK WAY" (SEE MAPS 1 AND 4)

Features: A scenic alternative to Route 160, but a little longer and slower. The gravel road is usually well graded but washboardy in spots. The road rises above the Cuchara Valley, providing nice views west and of the Wet Mountains to the north.

Approach: Travel west on Route 160 out of Walsenburg.

Mile 0: Turn left at the west edge of Walsenburg and enter 340 Road. You will be traveling southwest through the Bear Creek Valley rangeland. A few miles in on your right you will pass 342 Road, which wanders westward through a largely empty series of subdivided lots. Continue along the left fork.

Mile 7.9: Arrive at another Y-fork and bear right on 358 Road. Your route rises upward through pinyon-juniper country across a high glacial outwash terrace with commanding views in places across the Cuchara Valley to the north and west (Fig. 46).

Mile 16.7: Arrive on the eastern outskirts of La Veta: Moore Avenue at Walnut Street, the first stop sign in La Veta from the east.

DRIVE #1D: SULPHUR SPRINGS ROAD (SEE MAPS 1, 4, AND 5)

Features: A graded gravel road passes through flat open country southwest of La Veta, which affords good views of the Spanish Peaks from the northwest. The gently rising grade of the first several miles affords good views of Goemmer's Butte to the left, Raspberry Mountain straight ahead, and various tree-covered ridges. Then it rises more steeply into wooded country and a narrowing valley. Its destination is a low divide into the San Luis Valley. Beyond the Indian Creek Trailhead you need high ground clearance and dry weather.

Approach: Driving south of La Veta on Route 12, the road enters Route 12 from the west.

Mile 0.0: Beginning at the hardware store and golf course entrance at the south edge of La Veta, travel south on Route 12.

Mile 0.8: Turn right (west) onto 420 Road, which soon becomes 421 Road.

Fig. 46: A back road to La Veta from Walsenburg, looking across the plains toward the lower timberline of the P-J formation.

(The junction is 9.6 miles north of Cuchara.) A forest sign says "Indian Creek Trail 7 miles."

Mile 5.3: Sulphur Springs Ranch, a guest facility, has buildings on both sides of the road. A sulphur spring exists but is unseen and on private property. The manager reports that the once-hot spring has recently grown cold. To this point the road is standard two-lane gravel. Here, it becomes narrower with desultory maintenance, ad hoc applications of gravel or dirt, and roly-poly potholes—always a sign you have passed the driveway of the last taxpayer.

Mile 6.9: Indian Creek Trailhead. A small parking area on the left will accommodate six to eight cars (see Chapter 8, Trails). The road proceeds westward and steeply upward, passable only with high-wheel-base vehicles. One travels through the aspen and locust zone and into ponderosa pine and Gambel oak country. Nice sandstone rock formations occur on the right.

Mile 7.9: Entrance of side road FR410 on the right at elevation 9000 feet. FR410 is a dead-end road that enters the head of Tracy Canyon. It crosses open meadow with ponderosa pines and Engelmann spruce for about 1.5 miles. At the end, one can barely turn the car around. Avoid this road in wet weather. You will find a stream and nice rock ramparts at the end that are worth exploring on foot. The white firs, spruces, and ponderosas are quite large.

Passing the entrance to FR410, one continues on 421 Road. This road is only occasionally maintained and can be challenging. It crosses Indian Creek Pass, out of the national forest to the historic townsite of Fir, and onto the private Forbes Ranch properties of Costilla County. Check for conditions at the La Veta Visitor Center (Chamber of Commerce Office) and the Forest Service office in La Veta before attempting to go past the entrance of FR410.

CENTRAL

DRIVE #2: LA VETA–HUERFANO VALLEY LOOP VIA GARDNER, AND PASS CREEK PASS ROAD (SEE MAPS 1 AND 6)

Features: From the Cuchara River Valley, the route runs north into the Huerfano River Valley, then west up the Huerfano Valley, returning south up Pass Creek Road (sometimes called Pass Creek Pass Road) to Route 160 and back to La Veta. There are broad vistas in all directions, including a view over south Oak Canyon, which is the route of the Old Taos Trail over Sangre de Cristo Pass. The road up Pass Creek has attractive ramparts in spots and nice old cottonwood groves. Although Route 69 is paved, the county roads are two-lane gravel surfaces, washboardy in spots and only comfortable at a slow speed.

Approach: The route will be described counterclockwise from La Veta.

Mile 0.0: The junction of Route 160 and 520 Road just west and north of La Veta. Enter 520 Road, which goes straight north.

Mile 7.3: Pass the entrance of CR521, which enters on the right.

Mile 8.7: You are opposite the Black Hills on the east (right) and Silver Mountain on the left.

Mile 11.0: Cross a cattle gate and see a large rock on the right with faded historical notes. The entrance to CR530 Road is on the left (see Drive #2A, South Oak Canyon) (Fig. 47).

Mile 13.7: Route 69. Directly north of this intersection is the town site of Badito, the first county seat from 1866 to 1874 (see Chapter 3, History). What was left of it in 2001 is mostly collapsed today. Continuing west on Route 69 one passes through Farisita, now just a few buildings. Just beyond on the right is the pullout or overlook with the Arco Permian information panels that describe carbon dioxide extraction. To the southwest, you can look up South Oak Canyon toward Sangre de Cristo Pass. This is the Old Taos Trail route.

Mile 22.3: Gardner's Butte, a volcanic plug, is on the right.

Mile 23.3: The town of Gardner. On the west end where the highway makes a left jog is Kay's Country Store, purveyor of ice cream and some other food items.

Fig. 47: The Taos Trail historical plaque showing historic trails in the region. The plaque is on Yellowstone Road just south of Badito.

Mile 25.1: Turn left on CR550 toward Malachite.

Mile 30.3: Turn left at the sign indicating Pass Creek Road and Route 160 this way.

Mile 31.3: Arrive at the historic townsite of Malachite, where an old stone school still stands on the right (Fig. 10). Just beyond is Collector's Specialty Woods, a hardwood lumberyard and woodworker magnet. Bear left, then right, crossing the broad wetland of the Huerfano River. You have entered the Pass Creek Valley. Continuing southward across the narrowing landscape one enters a grove of large cottonwood trees. The road climbs close to smooth rocky ramparts on the left and emerges onto a broad, circular basin through which the road meanders. In the peak of summer, red penstemons and white evening primroses decorate the roadside. The route climbs more steeply toward the divide and emerges at Pass Creek Pass, overlooking Route 160 to the south.

Mile 41.3: Reach Route 160. A right turn on Route 160 would take you into the San Luis Valley and its attractions. To return to La Veta, turn left, cross North La Veta Pass (Fig. 48), and descend to the Cuchara Valley. Along the highway on the east side of the pass is a roadside pull-out area at Ojo Springs that emerges from Mount Maestas.

A monument honoring the war dead of Huerfano County and the memory of Private Felix B. Maestas, Jr., for whom the mountain was renamed, used

Fig. 48: The west-facing slope of Mount Maestas, shown here after an April snowfall, looms above North La Veta Pass on Route 160. Rough Mountain is to the left.

to be at that site. He was killed in action on the Italian Front in 1944. When members of his regiment were about to be overrun by advancing German troops, he covered their escape by killing 26 of the enemy from an exposed position and without regard for his own life. A few years ago, the vandalized monument was restored and placed in the Colorado State Veterans Nursing Home in Walsenburg. It was rededicated on Memorial Day 2003.

After reaching a low point approaching the La Veta turnoff, climb and watch for the sign for La Veta, a right turn at the top of the hill.

Mile 59.7: Reach Main Street in La Veta.

DRIVE #2A: ROAD TO SOUTH OAK CREEK CANYON (SEE MAP 6)

Features: Well-maintained gravel road. Wide open views in all directions. South Oak Canyon is probably the route of the Old Taos Trail. Most of the route and the pass are within private property.

Approach: From La Veta, head west briefly on Highway 160 and turn right (north) on 520 Road (mile 0.0).

Mile 11: There will be a left turn leading nearly due west, which is Road 530. At that turn, on the right side, will be a historical marker. (Arriving from the north, at the junction of Route 69 at Badito, the junction with Road 530 is 2.5 miles south.)

Take the left (west) turn at the cattle guard. The marker implies that this road is aligned with the old Taos Trail. There are majestic views west over a wide panorama. Shortly on the left is a photogenic weathered old church.

Mile 16.3: Reach a junction and turn right on 531 Road. (Continuing straight on 530 Road will take you to a private community protected by an extravagant and grandiose gate. It is a scenic area, but you will have to turn around.) Road 531 begins north, then climbs and continues west across a ridge.

Mile 17.4: Stop at the top of the ridge before it drops into South Oak Canyon (Fig. 49). The view to the west reveals South Oak Canyon to be a broad valley. At the summit of the valley is Sangre de Cristo Pass (private, with no public access). This is the valley and route over the Sangres followed by the Old Taos Trail. It is also, in my opinion, the route taken by the Anza forces in September 1779 after the Comanche defeat and death of Chief Greenhorn. This would have been the most direct route between Badito and the Fort Garland area. The passage of that expedition across the Huerfano River and over the pass was doubtless the county's most important historical event, at least in the 18th century. One can turn around here or, if you have the DeLorme

Fig. 49: Looking northeast over the Huerfano River Valley and Wet Mountains beyond. South Oak Canyon drops off to the left of this view.

Atlas or similar, continue north about 6 miles and emerge onto Route 69 near Farisita.

To do so, continue down into the valley and bear right on 531 Road, which follows South Oak Creek down canyon, eventually meeting 540 Road coming in from the left (west) at an intersection with a triangle of land in the center. Bear right on what has become 540 Road, which is better maintained, until reaching Route 69.

NORTH

DRIVE #3: GREENHORN PEAK AND WET MOUNTAINS LOOP (SEE MAPS 6 AND 7)

Features: This drive of about 137 miles can be enjoyed in one day from La Veta or Walsenburg. If you plan to hike much on Greenhorn or bag the peak, it is best to stay or camp in the Wet Mountain area and get an early start the next day. Views and weather are best in the morning. The road crosses extremely varied landscape from the plains to the alpine zone, from deep conifer forests to high-elevation meadows. Trails enter a dedicated wilderness area. The

U.S. Forest Service San Isabel Forest Map is useful for this area. In places, as described below, dry weather is required for safe passage.

Approach: The loop drive can be taken in either direction but, as described here, counterclockwise from Walsenburg seems a little easier.

Mile 0: From Walsenburg's north entrance at I-25, drive up I-25 to the Colorado City exit (mile 22) and take Route 165 west.

Mile 25.3: Greenhorn Meadows Park is on the left and contains a plaque commemorating the site where Comanche Chief Greenhorn was originally thought to have been killed (see Chapter 3, History, for a description of the battle and alternative opinion).

Mile 34.6: Enter the San Isabel National Forest. A steady upgrade passes road cuts that expose Precambrian feldspar and darker crystalline rocks. Pass the Custer County line and San Isabel Lakes Recreation District.

Mile 40.2: On the left is the little village of San Isabel.

Mile 43.6: Beulah Road (gravel road 78) enters on the right. Just beyond is the St. Charles Trailhead parking on the right (mile 44.4). Highway 165 continues its climb through good transitions of aspen, ponderosa pines, and spruces.

Mile 46.1: On the left is Bishop's Castle with parking on the right. This soaring facsimile of a knight's stone castle is open to the public and is worth a stop. The gift shop carries unusual medieval paraphernalia. As you enter the property you are confronted by large signs explaining how the owner's and our rights are being destroyed by our government. Without apparent irony, he then provides a long list of rules for visitors to accept—or no trespassing. Still, stop and stand in awe of this thing, a testament to the aesthetic obsession of a lifetime.

Mile 47: Turn left here on the sandy, gravelly road that enters the middle of a horseshoe bend. It is known variously as Forest Road 360 (or FR400 in older atlases) or Ophir Creek Road. Beaver dams are seen on the left and a beaver information sign. Creep ever upward through a forest of spruces, willows, occasional pines, and aspens.

Mile 54.1: At the top of a pass there is a four-way intersection and views over open meadows to the south. This illustrates the microclimate difference between the wetter north-facing and drier south-facing exposures. To the south is Road 634, which proceeds 21 miles straight ahead to its connection with Route 69 in Huerfano Valley. To the right is 401 Road to Deer Peak, 5.5 miles. Turn left on 369 Road that ends at Greenhorn Mountain parking area. Along the way there are several trailheads and parking areas.

Mile 63.3: Road cuts expose glacial moraine deposits: angular rocks in an earth mix. Gradually, the road passes upward through a timberline transition. It becomes rougher, rockier, and then single-track.

Mile 70.6: On the left is a primitive camping area and two lakes, apparently stocked, said a visitor.

Mile 71.2: Finally reach the parking area that has space for a dozen vehicles (elevation 11,456 feet, 37.8937°N, 105.0415°W). The northeast direction of the trail to the summit (elevation 12,347 feet) is obvious from the parking area. Views over the Huerfano and Wet Mountain valleys are spectacular.

A regular visitor claimed that the Wet Mountains really are wetter than Rockies ranges to the west. Warm, moist air rises off the Huerfano and Wet Mountain valleys and condenses over this range. This air is apparently warmer and more moisture-laden than that rising over the Sangres from the San Luis Valley (see Chapter 8, Trails, for selected descriptions).

Return to the four-way intersection.

Mile 87: Turn left (south) on 634 Road through spectacular open country with views of open range and distant mountains. The road is sandy and rough in spots. In places, when wet, the road turns to mud wallows. Use extreme caution. This is especially a problem driving uphill from the south.

Mile 91.4: The unusual white form of scarlet gilia (*Ipomopsis aggregata* subsp. *candida*) is found along the right for a couple of miles (Fig. 50).

Mile 93.6: On the left (east) is a side road to the Mission Wolf Sanctuary. There is a cattle gate, a small "M:W" sign with arrow, and a wolf with a raven on a router-cut sign. The sandy road is not stable when wet. Allow a couple of hours to see this impressive facility and these remarkable animals. People attempt to keep thousands of wolves and wolf-hybrids as pets in this country and then abandon or kill them when they refuse to domesticate. Here, they can live out their lives well cared for and in peace. We met founder and director Kent Weber, who gave us a good introduction to the site. Volunteers offer tours, and you will learn much about the history of wild and captive wolves in North America. The sanctuary is also a good place to see off-the-grid energy operations where they even generate enough solar voltage to use power tools. The volunteers are dedicated and interesting people.

Fig. 50: The white form of scarlet gilia (*Ipomopsis aggregata* subsp. *candida*) along the road between Gardner and the Wet Mountains.

Back at Road 634, continue downhill to its mouth.

Mile 107.9: Join Route 69 and turn left. A sign on a barn at that juncture says "We don't rent pigs." Continue southeast toward Gardner.

Mile 109.7: Arrive at the center of Gardner, Kay's Country Store, where you can get ice cream bars and some other groceries. Continuing east, a new grocery also sells ice cream.

Mile 119.3: Arrive at the Badito townsite on the left. Unfortunately, little is left of this, the first county seat of Huerfano County, which was much larger at that time. Just east, on the right (south), is the entrance to 520 Road toward La Veta. Or, continuing on Route 69 toward Walsenburg one passes about eight gob piles and old building foundations on the right. These are the remains of coal mining towns that flourished more than 50 years ago.

Mile 134.5: You reach the end of Route 69 where it intersects with the north drive out of Walsenburg. Turn right into town.

Mile 136.7: Reach the Safeway store on Route 160.

WEST

DRIVE #4A: MEDANO PASS ROAD (SEE MAPS 8 AND 9)

Features: The road crosses a pass in the northern Sangres and continues through pleasant montane scenery, some of heroic proportions. It has a little of everything. The eastern leg of this road is well-maintained gravel. This pass (or, as some say, Mosca Pass) was crossed by Zebulon Pike in 1806. Within the National Forest boundary, the road becomes mandatory 4WD, with big tires and high ground clearance being essential. At the end of your descent, within the national park, you are asked to remove some air from your tires for a better grip on a soft, sandy road section. Back on pavement, a free air pump station is provided near the visitor center. If uncertain about current conditions, call Great Sand Dunes National Park headquarters (see Resources) for their opinion.

Approach: The usual approach is to begin on the east. Proceed on Route 69 about 8 miles past Gardner (see Drive #2).

Mile 0: Turn left onto a gravel surface, clearly marked as the Medano Pass Road. This is 559 Road. It is a public thoroughfare passing through the private Wolf Springs Ranch. Regular big signs warn of dire legal consequences awaiting any hapless dimwit who leaves the road to trespass. You will proceed along a level or gentle upward grade.

Mile 2: You pass Creager Reservoir on the right and then begin climbing through ponderosa pine forest, followed by spruce-fir forest. With a sense of relief, you reach a sign declaring you are now within the San Isabel National Forest. Fine views are seen to the east and south.

Mile 7: A small campground appears on the left, which is also a good lunch spot. The serious (mandatory) 4WD road begins here. There will be some tortuous turns and boulders requiring full-concentration driving.

Mile 9: You reach Medano Pass (elevation 10,011 feet, 37.85583°N, 105.43194°W). The pass itself remains in timber with no immediate view. Take the time to park and walk up the trail following the ridge to the northwest for about 0.5 mile. You reach a turnaround and campfire ring. Great views loom to the east, south, and west.

Continue driving down the west slope. One-half mile down, a trail leaves on the right for Mount Herard, the peak that looms over the sand dunes to the north. About 0.3 mile farther is the first of several fords through attractive rocky creeks (elevation 9391 feet, 37.83639°N, 105.43806°W). Nice small campgrounds are found about every half mile on the way down. Upon reaching the final slopes, there are some fine bluffs arising right and left; then the sand dunes heave into view. Floral displays are not particularly diverse, but general scenery is excellent. At the canyon bottom you come upon a parking area where you are confronted with a 600-foot wall of sand, one of the region's great and inspiring sights. Giant cottonwoods frame the scene. Medano Creek runs along its base, except in severe drought years, having carved the canyon you just descended.

This is as close to the dunes as you can approach by road. To continue south, drivers are advised to lower tire air pressure to 15 to 20 pounds. One then meanders through a sand slalom for about a quarter mile before the roadway gets harder again. We found it not too challenging, but once underway, try to avoid spinning and don't stop on the sand! The late-afternoon sky provides good photogenic light on the dunes near headquarters. In general, if you desire good photo-ops of dune wind sculpture, lighting is best in early morning or late afternoon to dusk.

Near park headquarters is an air compressor facility where you should repressurize your tires. Returning to La Veta down the road south from park headquarters, turn left on Route 160.

Mile 45.4: Reach the town of Blanca. Continuing east takes you over La Veta Pass.

Mile 83.5: Brings you to La Veta.

DRIVE #4B: ROAD TO MOSCA PASS, AND TRAIL TO GREAT SAND DUNES NATIONAL PARK (SEE MAPS 9 AND 10)

Features: The road approaches from the east and climbs to the pass. A high-ground-clearance vehicle is recommended, but not necessarily 4WD.

If driving a regular car, be vigilant to avoid rocks that materialize here and there. From the pass, a foot trail drops downslope westward to the Great Sand Dunes. Once the principal access to the San Luis Valley, the trail was a toll road between 1871 and 1906.

Approach: (See Drive 2, La Veta to Huerfano Valley Loop.) From the east, the pass is achieved by driving a well-maintained gravel road, well-graded and certainly accessible by two-wheel drive. In a few spots it is definitely best to have a vehicle with higher ground clearance than the average car. As it climbs, the road is protected by guardrails in some places but exposed in others. Views to the east are excellent and display the valleys and surrounding peaks very well. Near the top, the road curves to the right and forks at the summit of the pass itself. A small sign reading "Great Sand Dunes Nature Preserve" indicates a left turn. Follow the dirt road a few hundred feet downward to the parking area and fence that stops further vehicular traffic. This road does not rise high enough to reach or approach timberline.

You will need your Colorado atlas. Just west of Gardner, a fork occurs where Route 69 turns northward.

Mile 0.0: Take the left turn, 550 Road, toward the west on the way to Red Wing. At the fork where 570 Road turns south is the Malachite townsite. Do not turn here. Continue southwest on 550, which becomes 580 Road at Red Wing.

Mile 7.0: Proceed to Sharpsdale (a garage-sized metal building these days). Bear right on 581 Road, which becomes 583 Road.

Mile 11.2: Arrive at the pass. This is the upper terminus of the Mosca Pass Trail (see Chapter 8, Trails). A sign near the fence notes "Sangre de Cristo wilderness, 226,455 acres."

If instead you follow the switchback that climbs to the right, the road goes a quarter mile farther and ends at a microwave radio relay tower and building. Don't bother stopping, as views are not good and the site is fenced, which inhibits exploring the area.

DRIVE #4C: UPPER HUERFANO BASIN: HUERFANO STATE WILDLIFE AREA AND LILY LAKE TRAILHEAD (SEE MAP 9)

Features: This gravel road meanders and climbs through a rugged valley and follows the Huerfano River toward its source. Hiking, fishing, and camping opportunities are available as one passes through the SWA. At the end is the access to the Lily Lake Basin, a spectacular hiking opportunity and one of the most rugged landscapes in the entire Rocky Mountains. A short walk west from your car yields a magnificent vista. Trailhead: 37.62360°N, 105.47295°W, elevation ca. 10,000 feet.

Mile 0.0: From Gardner (Kay's Country Store), proceed west out of town on Route 69.

Mile 0.7: Beyond the edge of town, turn left on 550 Road toward Red Wing.

Mile 5.2: Ignore the left turn to Malachite and Pass Creek Road, and continue westward. Upon reaching the metal building at Sharpsdale, the sign says "Mosca Pass, right, 6 miles." You turn left (southwest) where a sign says "Lily Lake 10 miles, forest boundary (upper Huerfano) 8.5 miles."

The next 10 miles to the trailhead for Lily Lake will take more than an hour to traverse. The road becomes increasingly intimate with narrow-leaf cottonwoods and willows on the roadsides. Within the SWA, there are numerous roadside pullouts, some with outhouses and space for camping and picnicking. They are popular with fishermen, hunters, and recreational campers. After you pass the last ranch entrance, road maintenance sharply deteriorates. Four-wheel drive is not required, but high ground clearance certainly is. Limited-slip differential is a plus. Heed the occasional "No Trespassing" signs that promise to prosecute. About 5 miles in, the views open up and subalpine meadows allow good vistas. The trailhead has ample parking space, but there are no visitor facilities (see Chapter 8, Trails).

DRIVE #4D: OLD LA VETA PASS ROAD* (SEE MAP 1)

Features: Originally this was a bed for a narrow-gauge railroad that began running in 1877, from Cuchara Valley to the San Luis Valley, once the highest railroad pass in the world. The last train passed this station in 1899, when standard-gauge tracks were routed through Fir, which goes over Veta Pass about 7 miles directly south. The route became the original Highway 160 alignment over the Sangre de Cristos, which was later replaced by the current alignment of Highway 160 over North La Veta Pass.

At Old La Veta Pass summit, you can see a few buildings from the once-functioning village. Continuing east you will find a combination of old pavement and gravel, which is usually well maintained.

---

*Also known as South La Veta Pass. The present Highway 160 goes over North La Veta Pass.

Approach: The road connects on the south side of Route 160 on both the east and west sides of North La Veta Pass. As described, the road is entered from the west, as the access is easier to locate from the highway. Therefore, the route is described counterclockwise.

Mile 0: Begin at La Veta Park and drive out using the west access road to Route 160. Drive west, cross La Veta Pass, and begin the long descent into the San Luis Valley. Pass the entrance to Pass Creek Road on the right.

Mile 16: A sign for Old La Veta Pass Road is placed on westbound lanes only. Turn left here. **Caution:** Be alert for traffic behind you when setting up your left turn, as many people barrel down this highway from the pass. The once-paved old pass road deteriorates frequently to gravel. The road now runs east and Mount Maestas appears straight ahead, with an abrupt gap on the left defining the edge of adjacent Rough Mountain. There are large aspens and extensive lodgepole pines, one of the few places where these pines can be seen from a road.

Mile 18.4: Summit of pass, 9380-foot elevation. This is also the boundary dividing Huerfano County to the east from Costilla County on the west. There may be a sign, "Uptop, pop. 2." The historic townsite, newly named by its owners, is being refurbished. At this writing it is mostly not open to the public. The small chapel of San Antonio on the right (south side) of the road was built in 1930 by lumber mill owner J.A. Trujillo and is open to visitors. Mass is celebrated in his honor each June 13th, the day of his birth and death.

The stucco building on the south side was the old railroad depot. On the north side of the road, the two-story building is the "Lone Pine Inn," which was run by a Walsenburg family and was famous for pies. A sawmill also operated that cut "pit props," or mine timbers. After the railroad's demise, the town was a truck stop on Route 160 until 1960, when the highway was moved north to its current alignment over North La Veta Pass (Fig. 51).

Continuing to head east, you will find fine views of the Spanish Peaks and adjacent valleys. The road descends.

Mile 26.1: The sharp right bend of Muleshoe Curve. Originally, its radius was thought by some to be too tight for trains, but it was successfully built and used anyway. Continue briefly uphill.

Mile 26.5: Reach the east entrance to Highway 160.

Mile 33.8: Reach the west entrance to La Veta off Highway 160.

Fig. 51: Descanso on Old La Veta Pass.

CHAPTER 8

TRAILS

EVERY LOVER of walking trails responds to the experience in a personal way. One may cherish the sense of aloneness, of closeness with nature, or of the sharp contrast between trail walking and life's ordinary distractions. I don't worry about spoiling this by encouraging too many people to walk trails. Too few will discover this delight. For the mathematically inclined, we had a saying in the park service, "The number of visitors to the backcountry decreases with the square of the distance from the highway, and with the cube of the altitude." This seems to remain true, and this corner of Colorado remains as unspoiled as it gets.

GENERAL CONSIDERATIONS

For those new to trail hiking, please see Appendix 1, which goes into some detail about conditioning and selection of gear. While some of the trails are open to motorized vehicles, I am stressing muscle-powered recreation in this chapter. What follows is a brief list of considerations:

1. **Trail ratings vs. your condition:** I've chosen not to offer any systematic difficulty ratings. Concepts of what is easy or difficult vary with the individual. I will note subjectively the nature of the trail with each description. You need to know your limits and be conservative in estimating your capability. Allow enough time, and don't be embarrassed to turn back. Some of us were raised to think we always have to reach our "objective" no matter what. Where is that chiseled in stone? The trail will be there next week! Also, a short, slow walk affords many pleasures. You will notice more of nature's beautiful details than fast-moving "power-hikers."

2. **Safety:** Be sure that others know your route, destination, and approximate time of return. This cannot be overemphasized. No matter how long your walk, carry water and some extra food beyond your expected needs.

3. **Weather conditions:** Unless conditions seem immediately life threatening, I always go on a trip as planned. Conditions may change dramatically, giving you some terrific atmospherics or lighting conditions for photo-ops and just plain enjoyment. The key is to dress for and carry gear appropriate to the season, and you can be quite comfortable. But consider lightning. In simplest terms, stay off high, treeless (alpine) elevations in the afternoon, especially in late summer. Throughout the West, mid- to late-summer afternoon cumulus cloud buildup makes the high country very dangerous. All protruding objects (including yourself!) are vulnerable.

4. **Trail etiquette:** If you must bring pets along on a hike, keep them leashed. Although some are well behaved, others disturb wildlife greatly, not to mention those with whom you share the trail that day. For certain you will see no wildlife, which otherwise is a joy of being in the backcountry.

Leave your radios and tunes at home, unless you wear earphones. But think about why you are out there. "Climb the mountains and get their glad tidings," as John Muir once said. Nature in itself has much to offer if you give it a chance.

Keep things picked up. Litter from your lunch, snacks, film wrappings, and so on is offensive to other hikers, degrades the environment, and is not good for wildlife. Carry a bag to pack your trash and that of others when possible.

5. **Observing wildlife:** Walkers serious about bird watching or animal tracking go singly or in very small groups. They talk little and in low tones. Early and late in the day are good times, especially around water sources. Conversations while walking are terrific for visiting and companionship, but don't expect to see many other animals.

TRAIL LIST

From Cuchara, originating near Route 12:
Indian Creek Trail—north end
Dike Trail
Dodgeton Creek Trail
Baker Creek Trail
Boyd Mountain

From Blue and Bear lakes region:
North Fork Divide
Indian Creek Trail—south end

Teddy's Peak
Trinchera Peak

Trails on the Spanish Peaks—north side:
Bulls Eye Mine Trail
East Spanish Peak

Spanish Peaks—south side:
Cordova Pass to timberline
West Spanish Peak
Apishapa Trail
Shafer Trail
North Fork Trail—south end

Western Huerfano County:
Mosca Pass Trail
Lily Lake—Upper Huerfano Basin

North:
Greenhorn Mountain

East, Plains:
Dinosaur track site

TRAILS FROM THE CUCHARA VALLEY, ORIGINATING NEAR ROUTE 12

INDIAN CREEK TRAIL—NORTH END (SEE MAPS 5, 11, AND 12)

Trailhead: On Sulphur Springs Road at mile 6.9 from Route 12. National Forest sign and parking area for several cars on the left (south).

Quadrangles: McCarty Park; Trinchera Peak 1:24,000.

Approach: (See Chapter 7, Drive #1D, Sulphur Springs Road.) From Route 12, take Sulphur Springs Road (421.1 Road) west, opposite Grandote Golf Course. The Sulphur Springs Road junction with Route 12 is 9.6 miles north of Cuchara and about 1 mile south of La Veta.

Main Theme: A steady climb, occasionally steep, for several miles up the north side of Raspberry Mountain. The elevation gain from the trailhead to the summit of Raspberry Mountain (8440 to 11,241 feet) makes this a potentially challenging hike. Various vegetation changes and views occur as one ascends. It can be enjoyed as a nice out-and-back walk or as a connection to Bear Lake. One can descend the Dodgeton or Baker Creek trails before reaching Bear Lake. You will need topographical maps to find these connections, as they may not be marked. For longer routes, arrange a shuttle with a friend by leaving one vehicle at the other end.

Description: After about a quarter mile, views of forest to the south and east begin to open up. The trail proceeds into a fine mature forest of ponderosa pine parkland. It then adds white fir and mature aspens with an understory of Gambel oak and snowberry. August flowers are yarrow and monkshood; near the stream are two species of purple asters, scarlet gilia, and various yellow members of the aster family. Picturesque sandstone formations along the trail punctuate the landscape.

Here and there a sill (horizontal dike) has pushed into the country rock. Along the trail the sill outcrops at about 4 feet high along with evidence that the sedimentary rock was baked (heat-altered) at its margins (Fig. 52). After proceeding steadily uphill for about 3 miles, one finds the marked boundary of the Forbes Ranch properties to the west. The trail follows the ridge with gentle ups and downs to the 4-mile marker. (Mile markers are on low posts and are often obscure.) At several points, one can walk west a short distance for views

Fig. 52: Outcrops of the Sangre de Cristo Formation along the north end of the Indian Creek Trail.

northwest to Mount Maestas at the Forbes Ranch boundary. Shortly after the 4-mile marker the trail becomes steep and proceeds to a ridge apex or saddle.

For those who want to see the summit of Raspberry Mountain, low vegetation allows one to see the direction to bushwhack northeast to the north summit. This is about 1000 feet for a 150-foot elevation gain. Although the summit is bald, there is no view due to surrounding Engelmann spruces, aspens, and subalpine firs. There is a picnic fireplace and two rock cairns. Returning to the trail and proceeding south for about three fourths of a mile along the summit ridge, one could bushwhack to the south summit. If you continue south on the trail for another quarter mile, there are good views of the entire Cuchara Valley to the south.

For the entire distance to Bear Lake, about 14 miles, the trail is passable by four-wheelers and is approved for that use. These vehicles have beaten up the trail, causing deep erosion in several places. Still, this trail remains an interesting walk on the west side of the Cuchara Valley.

DIKE TRAIL (SEE MAP 13)

Trailhead: Elevation 9020 feet, 37.35286°N, 105.10071°W.

Quadrangle: Cuchara Pass 1967, 1979. 1:24,000. But note that upper and lower trail approaches are improperly located.

Main Theme: "Dike" is a misnomer as the trail climbs to and follows the Dakota Wall, an ancient Cretaceous sandstone beach. Despite appearances, it is not one of the hundreds of true dikes that radiate from the Spanish Peaks. The trail runs through a varied and complex set of ecological situations ranging from a mature and majestic conifer forest through young aspen, aspen jackstraws being replaced by firs and ponderosa pines of all ages, to overgrown scrub.

Description: At the south end, the trail runs from a small parking area on Park Road in the Spanish Peaks Subdivision to a small parking area in the Cuchara Subdivision. From the south, the trail runs up a steep slope in a series of 15 switchbacks, which are well laid out and maintained. It then follows along the base of the Dakota Wall until it descends to the town of Cuchara.

From the south end, the lower switchbacks rise gradually through mature ponderosa forest parkland (Fig. 53), with specimens as large as those found on any public trails in this region. Open vistas give pleasant views west and south with shrubby ground cover featuring mountain snowberry, mountain mahogany, and Gambel oak. Other large trees are white fir, including two of the largest mature specimens to be found in the region.

Fig. 53: Mature ponderosa pine (*Pinus ponderosa*) on the Dike Trail.

Higher up one enters a denser forest of younger white fir that is regenerating well within a dying aspen woods. The south ends of the last two switchbacks have short spur trails, providing nice vistas of Cuchara Pass and the southern end of the valley (Fig. 54). At a sharp left turn (walking from the south), a short, steep trail ascends to the east to an overlook at a low spot in the Dakota Wall. Fine views of the White Peaks can be seen (Fig. 55).

The northern two thirds of the trail is quite variable. For a while it follows an old road, then it seems to reduce to an overgrown game trail in places. The excellent maintenance on the south end becomes desultory on the north. Planning and construction are poor. Places that need switchbacks lack them. Some maintainer slashed off fir branches with a chainsaw, leaving large scars on the trunk.

In the dense vegetation of the northern portion one might see a blue grouse explode from the thicket. Plants seen on the north end include meadow rue, *Prunus* L. species, weedy grasses, *Rubus* L., ponderosa pine, white fir, Douglas fir, common juniper, *Gaultheria* L., cow parsnip, several species of *Mertensia* Roth and *Cirsium* Mill., plus black locust (small- to medium-sized trees), several Apiaceae, mountain snowberry, creeping mahonia, and cardinal lobelia, which is said not to occur above 5000 feet.

DODGETON CREEK TRAIL (SEE MAP 12)

Trailhead: Elevation 8585 feet, 37.37222°N, 105.10556°W.

Quadrangles: Cuchara and Cuchara Pass 1963, 1979. 1:24,000.

Approach: The trail begins at a parking area just off Route 12. Just south of Cuchara, turn west (right) at the sanitation and water district office driveway and proceed uphill about 200 feet to the official U.S. Forest Service (USFS)

Fig. 54: Cuchara Pass on the left horizon from the Dike Trail. An unnamed peak arises on the right side of the pass.

Fig. 55: Overlook to the east on the Dike Trail, looking northeast toward the rock glacier on North White Peak.

Fig. 56: A typically solitary ponderosa pine on the Dodgeton Creek Trail. Such giants are vulnerable to lightning strikes.

Trailhead, where a concessionaire collects $5.00 if you park your car there.*

This begins as the "Spring Creek Trail" but quickly climbs out of the Spring Creek Valley, then enters the Dodgeton Creek drainage and is called by that name the rest of the way. (The name is a corruption of the original settler's name, Dogion.)

Main Theme: The trail is wide with open vistas and frequent spectacular views. There is lots of variety. It is an easy-walking, well-packed dirt trail with well-designed switchbacks most of the way.

Description: The trail begins in an old spruce fir forest, then climbs into open ponderosa pine parkland (Fig. 56) with a Gambel oak understory. The views east really open up after the first quarter mile: the three White Peaks and West Spanish Peak fill the panorama. There are patches of aspen, a few Douglas fir, and frequent patches of low-growing common juniper. Dwarf mistletoe sprouts from branches of ponderosa pine. The trail proceeds in a westward direction and then turns south. A white fir stand is well established, replacing an old aspen forest. After about 40 minutes, a large meadow allows views in all directions (37.36833°N, 105.11917°W). Just past the meadow a trail junction at 1.3 miles allows a link to the Baker Creek Trail to the south. Indian Creek Trail can be reached in 3.1 more miles. (A topo map is valuable here, as the network of trails can be confusing.)

A second meadow is reached after a quarter mile, and the grade becomes less of a climb. Lodgepole pines become noticeable, and soon many pure stands and aspen-mixed stands become dominant. Alders, eastern scouring rush (*Equi-*

*Recreation user fees (basically for parking) began experimentally in 1996, but the Bush administration moved to make them permanent. On the argument of double taxation, they are subject to ongoing litigation. A recent federal district court decision has upheld their legality. Holders of Golden Age or Golden Access passes can get a 50% discount at a Forest Service Office.

setum hyemale), and tall yellow-green *Pedicularis* L. (Indian soldiers) are common. Just beyond a ridgetop, a fork in the trail is unmarked and confusing. To the left a dominant path goes uphill; to the right a path goes downhill. The right path is the Dodgeton Trail. If you take the left path, you soon reach a picnic site with a pole pine kitchen table complete with a cracked frying pan hanging on a tree. Once used by Cuchara Recreation Center outings, it remains a nice place if it's lunchtime. From there a nearly invisible trail climbs uphill and joins the Indian Creek Trail. Go right to follow the Indian Creek Trail north.

If you mean to climb toward Raspberry Mountain, that option will cost you 1 mile of progress, so it's best to skip the picnic table.

At the correct Dodgeton Trail junction with the Indian Creek Trail (elevation 9959 feet, 37.38750°N, 105.14694°W), there are signs pointing north (FDR, access road 7.1 miles) and south (Bear Lake campground 5.6 miles).

Indian Creek Trail can be followed fairly level for miles at this point by walking down an endless road through unexciting lodgepole pine stands. The understory plants beneath these stands range from depauperate to absent. Their age and condition indicate that no fire has been in the area for more than a century.

BAKER CREEK TRAIL (SEE MAP 11)

Trailhead: Elevation 9491 feet, 37.36056°N, 105.12722°W.

Quadrangles: Cuchara Pass 1967, 1979, and Trinchera Peak 1994, both 1:24,000.

Approach: The trail is reached by the Cuchara Mountain Resort Road to the (west) right off Highway 12 south of Cuchara. At the end of the road, after a distance of 1.3 miles, one can park near the large blue water tower. Halfway up, one reaches a Y-fork in the road opposite the main resort buildings. Take the left. The parking area for three to four cars by the blue water tower is the trailhead. Do not park elsewhere, as you will be politely reminded by the resort security staff that you are trespassing. There is a modestly marked USFS sign saying "Baker Creek Trail." Too frequently locals leave construction trash in the parking area.

Main Theme: This trail is relatively little used, narrow, and intimate, with few views and lots of wildflowers. It gives a close-up view of nature. Few area residents know it exists.

Description: Baker Creek Canyon Trail once began at the edge of Route 12 but was obliterated by development of the ski resort in 1981, a removal of land from the national forest that occurred under mysterious circumstances.

From the current trailhead, the first 200 to 300 feet of trail is narrow and rises through an old aspen grove nicely infiltrated by small- to medium-sized Engelmann spruces. In June, look for golden banner (Fig. 57), wild geranium, vetches, orange Indian paintbrush, mountain maple, and box elder. The trail ends in a T-intersection, which is the junction of two trails.

To the left (south), the trail heads south and then west to connect to the Indian Creek Trail at the foot of Steep Mountain, after several switchbacks. On that trail, one finds the common understory shrubs, mountain snowberry, mountain mahogany, and common juniper. Gambel oak occurs on exposed areas around open meadows with views. Flowering plants include Rocky Mountain locoweed, gold banner, and hound's tongue. Nice views to one's left frame West Spanish Peak and the south end of Cuchara Valley (Fig. 58). In the near ground is the defunct ski resort, an installation whose permit to operate was removed by the USFS in 2002, two years after it permanently ceased operation. Regulations mandate that it be restored to forest. But there it sits in agitated stagnancy waiting for an eighth marriage (business) proposal—the triumph of hope over experience?

Meanwhile, 10 minutes south of the fork, the trail curves to the west and enters a valley and follows Baker Creek. Be on the lookout for animals here, or whenever there is creek noise. Animals cannot hear your approach well enough to be out of your way before you see them. You will find mountain maple,

Fig. 57: Golden banner, a non-native legume, adds bright yellow blossoms to the montane landscape.

Fig. 58: View of West Spanish Peak from an opening in the Baker Creek Trail, just north of its connection with the Indian Creek Trail.

hound's tongue, wild rose, heartleaf arnica, cow parsnip, and *Lonicera involucrata* (involucred honeysuckle).

The forest is mostly Engelmann spruce with an occasional white fir above 9461 feet. Watch for one nice glimpse of West Spanish Peak. Other flowers include bluebells, elderberry, horsetail, and false Solomon's seal. In the mature forest, look for coralroot orchid, rattlesnake plantain orchid, and red columbine. After a section of steepening trail and several switchbacks, the Indian Creek Trail junction is reached at 10,461 feet. It is mostly level at that location. The time required to reach the Indian Creek junction is approximately 1.5 hours.

Returning to the junction just above the Baker Creek Trailhead, the trail to the right (north) connects to the Dodgeton Creek Trail. It climbs into an open area with nice views. The same plants as above can be found in addition to some bristlecone pines, limber pines, mountain maple wild rose, hound's tongue, heartleaf arnica, and cow parsnip.

BOYD MOUNTAIN (SEE MAP 13)

Trailhead: There is no trailhead or public parking area on the north end. On the south end, one can park at one of the small roadside parking areas on the lower Bear Lake Road. This is a cross-country scramble from about 9000 feet with an elevation gain of 1500 feet. Time to summit: about 2 hours. This description will begin on the north end.

Quadrangle: Cuchara Pass 1:24,000.

Approach: Off Route 12, west side, opposite Spanish Peaks Subdivision. Hiking may begin anywhere from the area of the Cuchara Chapel, at the entrance to the Cuchara Mountain Resort, or the south (Aspen Road) entrance to the Spanish Peaks Subdivision.

Main Theme: Boyd Mountain is the large, though not high, mass to the west of Route 12, between the Cuchara Mountain Resort Road on its north end and the Bear Lake Road on its south end. Although lower than the crest of the Sangres, nice views are afforded in various directions on the summit ridge. This is a steep mountain, and there is no easy path.

Description: Leaving from the Spanish Peaks Subdivision south entrance (elevation 9008 feet, 37.34750°N, 105.10139°W) bearing west, walk up the northeast-facing slope, which is about a 24% grade. One ascends through a diverse open forest composed of mature second growth of Douglas fir, limber pine, ponderosa pine, aspen, white fir, and Engelmann spruce. The aspens frequently have claw marks and antler rub marks, and there are frequent bear and elk droppings. In October, the bugling of elk in rut can be heard across the valley. The ground cover of blueberries is heavily browsed.

Upon reaching the summit ridge, note that the vegetation becomes somewhat stunted because of more severe exposure. Proceed southward until gaining the summit meadow at the south end of the mountain. There are faint game trails in spots, but keep climbing. Summit position: elevation 10,523 feet, 37.33500°N, 105.17111°W. Along the way are nice views of the Sangres' crest to the west. There is a fire ring, probably used by hunters, and a 5-foot-high tepee of branches (Fig. 59). Nice views exist of Cuchara Pass, the upper Cuchara River Valley, and Bear Lake Road. White firs, limber pines, moun-

Fig. 59: Looking east from the summit of Boyd Mountain, toward West Spanish Peak.

tain mahogany, and common junipers ring the summit meadow. On the west side are aspen, white fir, juniper, bearberry, Engelmann spruce, and pine drops. Descend to the west with the Bear Lake Road in view below. At the saddle to the west (elevation 9955 feet, 37.32917°N, 106.11417°W), one encounters a very old stand of tall white firs and limber pines. One then descends gradually, along any gradient to the road. It is simple to return along the highway, or through the subdivision along Aspen Road.

TRAILS REACHED FROM THE BLUE AND BEAR LAKES REGION

NORTH FORK DIVIDE (SEE MAP 14)

Trailhead: Elevation 10,750 feet, 37.30972°N, 105.13944°W.

Quadrangles: Cuchara Pass and Stonewall 1967, 1979. 1:24,000.

Approach: Take the road to Bear Lake and Blue Lake (to the west) that leaves Route 12 in Cuchara Valley. Just two switchbacks past Blue Lake, the road bears left just before another switchback. This road becomes a four-wheel drive (4WD) road to Trinchera Peak. Parking is available on the right just inside this road entrance.

Main Theme: For the first mile, one walks upward through a mature stand of Engelmann spruce that ends at an alpine meadow with spectacular views in all directions, including several of the Sangre de Cristo Range's best-known peaks. Wildflower displays are diverse and excellent, at this writing, in a year of the worst drought in living memory.

Description: The trail rises about 1000 feet during a leisurely 3½-hour round-trip that includes lunch. One walks up the Trinchera 4WD road, which climbs steadily upward through several switchbacks for about 1 mile (about 1 hour). Along the way is an understory of low-bush blueberry. In openings, look for stands of larkspur, blue columbine, alpine clover, chrysanthemum, and geranium. Lots of light-green old man's beard lichen festoons spruce trunks and branches (Fig. 60).

One approaches a trail merging on the left whose sign reads "North Fork Trail, Purgatory campground 5 miles." Take this trail. (The other trail goes toward Trinchera Peak.) No longer a road, this walking trail climbs gently before soon dropping over the divide into the Purgatoire Valley, the headwaters of the Purgatoire River. That trail (which continues straight ahead) continues downhill into the valley. (Beginning at the downhill end, this trail is discussed separately as beginning from the Purgatoire campground. It climbs into beautiful meadows.)

Fig. 60 (left): Old man's beard lichens (*Usnea* species) growing on spruce bark above 11,000-foot elevation.

Fig. 61 (below): Harebells (*Campanula rotundifolia*) in North Fork meadow.

Soon a substantial unmarked trail, known by some as the "Wildcat Trail," appears on the left. Take that trail. It angles gently upward through thinning timber and fine wildflower displays. Look for mule ear daisies, wallflowers, harebells (Fig. 61), marsh marigold (Fig. 62), pussy toes, bistort, and mountain clover.

The view gets progressively more spectacular, and soon the entire Culebra Range heaves into view to the west. You are walking south and ascending the east ridge of the Purgatoire North Fork Valley. Continue on until you reach the ridge top to your left. Bristlecone pines appear, as well as Engelmann spruce trees that are pruned relentlessly by the winter winds into severely one-sided krummholz. Higher, numerous clumps of dwarfed Bebb's willow occur as well as numerous logs and standing dead trunks, victims of a forest fire in this area, perhaps a century or more ago. Reflect on the time it takes to recreate a forest once destroyed on sites like this one. Today, standing and fallen wood shows a singular patina produced by years of sun bleaching and sand polishing by the wind.

Occasionally, one may see herds of elk crossing the ridge and certainly evidence of where they've been in this obviously favorite meadow.

On the ridge top (elevation 11,770 feet, 37.29639°N, 105.13944°W) is a good place for lunch. One has a commanding view of Goemmer's Butte and Greenhorn Peak to the north, the Spanish Peaks to the east, Fisher's Peak down the Purgatoire Valley to the southeast, and most of the Culebra Range from Teddy's Peak to the north to Culebra in the south, the southernmost fourteener in Colorado.

INDIAN CREEK TRAIL—SOUTH END (SEE MAPS 11 AND 14)

Trailhead: Walk northward from the Bear Lake campground to a trail running north and a sign for Indian Creek Trail.

Quadrangle: Trinchera Peak 1:24,000.

Approach: Arrive at Bear Lake campground from the Blue Lake–Bear Lake Road off Route 12. The parking space is a fee area.

Main Theme: This is the south terminus of a 14-mile trail whose north end joins the Sulphur Springs Road parking area. It runs along the base of Teddy's Peak, over Raspberry Mountain, then descends to its northern terminus.

Description: The well-designed trail runs uphill to a saddle and a fence alignment. It continues northward at that point, 14 miles to the terminus on the Sulphur Springs Road. To the right is a narrow trail that runs to the saddle overlooking the Cuchara Valley and ends near the top of the abandoned ski run. Very nice views here show the Spanish Peaks, the White Peaks, Cuchara Pass to the south, and the Huerfano Valley to the north (Fig. 63). Young bristlecone

Fig. 62: Marsh marigold (*Caltha leptosepala*) cluster on the North Fork Divide. It is common in high-elevation wetlands.

Fig. 63: West Spanish Peak from Bear Lake Overlook. Multiple cloud layers are common in the fall.

pines, Engelmann spruce, and recumbent junipers are common in the area. Early season wildflowers are cinquefoil, paintbrush, arnica, pussy toes, yarrow, wild buckwheat, sedum, harebell, pink clover, groundsel, asters, and wild strawberry. This is about a 1.5-hour round-trip from Bear Lake campground.

TEDDY'S PEAK (SEE MAPS 11 AND 14)

Trailhead: Elevation 10,748 feet, 37.32467°N, 105.14972°W.

Quadrangle: Trinchera Peak 1994. 1:24,000.

Approach: One approaches by driving around Blue Lake where the road curves to the northwest and ends at Bear Lake. Halfway to Bear Lake, look for the side road on the left (west), FP 413. Drive to its end and park. This is the jumping-off point rather than "trailhead."

Main Theme: No actual trail exists. This is a cross-country wilderness jaunt, not to be attempted alone by novice hikers (see Appendix 1, Preparing for Outings). Carrying and knowing how to use a map and compass or a GPS device is required. The rewards are majestic forests, a classic stunted timberline, wildflowers, a walk over exposed rocks, and magnificent views. The climb takes about 3.5 to 4.0 hours. The destination, Teddy's Peak, is at an elevation of 12,581 feet (37.34111°N, 105.16639°W). There are fine views in all directions.

Description: Proceed on a compass bearing of 230 degrees up the slope to the west through an Engelmann spruce forest. As the trees become dwarfed, follow a bearing of 270 degrees. Alternatively, one could follow a trail near the stream that parallels your course to the right (north), but this leaves you boxed into too steep a place to make an efficient assault on the summit. It is best to follow the main slope, always upward, and more or less parallel to the stream to the right. (Near forks of the stream: elevation 11,286 feet, 37.32306°N, 105.15667°W.)

The forest is very nice and not hard to travel through. It is old growth in spots, and wind-thrown trunks may slow your progress in places. Tree diameters range from 2 to 4 feet, with heights up to 150 feet near the stream. Pit-and-mound topography in places is not hard to navigate around. The understory is mostly lowbush blueberry, *Arnica* L., *Hydrophyllum* L., *Pedicularis*, purple aster, and the attractive *Moneses uniflora* (Pyrolaceae).

One emerges into the sunlight in an elfin forest, then timberline on this ridge (elevation 11,640 feet, 37.32250°N, 105.16056°W). There are cinquefoil, *Heterotheca* Cass., blue columbine, harebell, and wild buckwheat.

One then angles northwest toward the crest of the main treeless ridge, which is easily seen from here. Cross a shallow, dry valley into a nice middle-aged, gnarled stand of bristlecone pines. This is a good place to stop for lunch or a snack. None of the bristlecone stands are really old on this range, compared with the ancient giants on Mount Evans to the north, at the lower edge of the Front Range.

From the pines one walks above timberline northward to the summit of Teddy's Peak. As with many peaks, there is a summit cairn, which, in this case, is a low semicircular wall that provides protection from the wind while you eat, hydrate, and rest. Is this an ancient Indian surveillance structure? I have no confirmation yet, but it seems possible. Tucked under a rock you will find a jar with a small notebook: the summit register. Sign in and note how few people reach this point in a given year.

Returning, one can descend between the creek forks through a mature spruce forest. This is easy walking, with huge trees, but also with ever-increasing steepness. At the creek junction, one finds a game trail that follows the main stream down to the trailhead (car park). Deep, well-watered flower beds display bluebells, larkspur, and groundsel. One could make an argument for proposing a trail here, but it is really nice as it is.

TRINCHERA BASIN AND PEAK (SEE MAPS 11 AND 14)

Trailhead: Elevation 10,878 feet, 37.31017°N, 105.13894°W.

Quadrangle: Trinchera Peak 1994. 1:24,000.

Approach: Take the drive to Blue Lake and Bear Lake from Route 12 south of Cuchara (see Chapter 7, Drives). After arriving at the Blue Lake parking area, continue upward a quarter mile through two switchbacks. Look for a side road to the south that leads to Purgatoire Divide and Trinchera Peak. Take this side road about 0.1 mile to a parking area on the right, just out of sight of Blue Lake. Parking is at 37.31017°N, 105.13894°W, elevation 10,878 feet.

Main Theme: A walk or drive up an old mining road. If driving, 4WD and high ground clearance is required. The trail (road) leads through a spruce forest of variable age, opening out into a large tundra habitat. Sightings of bighorn sheep and marmots are likely, and there are good wildflower displays in midsummer at high elevations. Trinchera Peak itself is a major summit of the southern Sangres with broad views (Fig. 64).

Description: After about 50 minutes of walking, and about nine switchbacks, you reach a trail junction and sign: Trail #1309. The North Fork Trail bears to the left (southeast). It ends at the North Fork campground in 5 miles. Trinchera is 2 miles onward to the right (southwest). Take that trail.

The road continues gently upward and is curvy but has no switchbacks. Many of the trees are venerable but not large. One core sample of a 1-foot-

Fig. 64: Looking north from the upper slopes of Trinchera Peak, from just above the saddle at the end of the old miner's road, seen in the middle ground.

diameter spruce showed that it was about 100 years old at the time of the American Revolution!

About a half hour above the junction you reach timberline rather abruptly. You can look south and west over a couple of thousand acres of willow thickets. As you continue up the winding and rocky road, the willows become more stunted and mixed with wildflowers of many species. Broad views of the Spanish Peaks and plains beyond are spectacular. The road reaches the north face of Trinchera (37.29502°N, 105.16000°W, elevation 12,191 feet) about 1.5 hours from the North Fork junction.

In good snow years, a large snow field persists there into mid-July. This basin is a favorite lunch spot. Bighorn sheep may examine you from over the ridge or across the snow field. Marmots often sun themselves at a distance, as they are not accustomed to people.

Continuing upward, the road ends at the saddle (elevation ca. 12,500 feet), where you look steeply down onto Forbes Ranch properties to the west. You can also walk south to the old mine site.

From the saddle, one can climb to the summit of Trinchera Peak (elevation 13,517 feet) via its north ridge and enjoy sweeping views of the San Luis Valley to the west, New Mexico to the south, and the Spanish Peaks to the east. The narrow, rocky route passes over a notch and then up to the main peak. Although it is steep and strenuous, it is one of the most rewarding climbs of the area. The summit marks the southwest corner of Huerfano County at its boundary with Costilla County on the west and south. The Las Animas County line is about a half mile to the southeast.

TRAILS ON THE SPANISH PEAKS—NORTH SIDE

Bulls Eye Mine Trail (See Map 15)

Trailhead: Elevation 8428 feet, 37.42250°N, 104.97639°W.

Quadrangle: Spanish Peaks 1971. 1:24,000.

Approach: The Wahatoya Road goes south out of La Veta and arrives at a small summer cabin community in a steep valley. On the western edge of the cabin valley, you reach a ridge with a parking area just beyond the summit. The trail leaves from the ridge top immediately west of the parking area.

Main Theme: This trail runs steadily upward following an old roadbed much of the way. It climbs more steeply than a well-designed trail and is a real test of your condition. After some elevation is gained, there are good views of East Spanish Peak to the east. Views from the top above timberline in the cirque basin are fine to the north. The round-trip is approximately 7 miles.

Description: Elevational gain is about 3000 feet in 3.5 miles. At the forks of the Apishapa Trail (37.39722°N, 104.97444°W, elevation 9969 feet), which takes off to the left, continue straight ahead.

The last grove of trees opens out onto a huge talus field (37.38972°N, 104.97972°W) and makes a good lunch stop. The spruce-fir forest ends abruptly at the edge of a classic alpine zone. Good populations of penstemon flowers decorate the trail in late July. From that spot to the mine site, at 11,200 feet on the west wall of a steep cirque basin, is another 1.5-hour round-trip. What was once a deep cirque lake had been emptied by a fast-eroding V-outlet in the terminal moraine. A creek continues to drain a melting snow field.

Once owned by Robert Lincoln, the president's son, the former mine opening is a small depression in the ground with a few scattered mine timbers. Its position high on the cirque wall is spectacular. On a clear day, one can see past the towns of La Veta and Walsenburg, out onto the Great Plains beyond the Wet Mountains (Fig. 65).

EAST SPANISH PEAK (SEE MAP 15)

Trailhead: At the end of the Wahatoya Camps Road. Elevation 8600 feet, 37.42250°N, 104.97639°W.

Quadrangle: Spanish Peaks 1971. 1:24,000.

Approach: An approach from the west works well using the Wahatoya Road running north out of La Veta. The coordinates given are for the Bulls Eye Mine Trailhead. Continue along Wahatoya Road about ¾ mile to the east end of the Wahatoya camps. There is enough space for one or two cars.

Main Theme: Although technically classed as a "walk-up," this is a steep and very challenging climb. Some experienced climbers consider this the most difficult "twelver" (peaks over 12,000 feet) in Colorado, a category with lots of candidates in the state. The climb penetrates deeply into the newly designated Spanish Peaks Wilderness Area. Scenery is majestic most of the way, and substantial portions of your effort will be above timberline. Watch the weather. Because of lightning potential, do not walk into dark clouds, and get an early start.

Fig. 65: Looking north from the Bulls Eye Mine site on West Spanish Peak. Walsenburg and the edge of the Great Plains are visible to the right. The elevation in the center distance is Greenhorn Peak and the Wet Mountains.

Description: While its summit is 1000 feet lower than West Peak, the climb's difficulty is due to beginning at elevation 8600 feet, thereby requiring a 4000-foot gain from the nearest approach trailhead. The round-trip is about 10 hours. A dawn start is advisable, especially late in the season. The atmosphere and views are better earlier in the day, but, more importantly, one needs to be off the summit during the afternoon rain and lightning.

A good trail runs along Wahatoya Creek for about ¾ mile. Looking at the map, one might think to attack the ridge on the left (north), but let the feeling pass. The canyon walls are too steep and there are several unproductive false summits.

When the creek trail ends, proceed up the ridge directly to the south (compass bearing 190 degrees). At about 10,280 feet, one intersects the pack trail, a good trail that roughly follows the contour going east. It crosses the Las Animas County line, marked by a sign, at a divide up the west saddle ridge. A small trail marked by plastic flagging, then small rock cairns (unmapped), follows the county line all the way up the southwest ridge to the summit at 12,683 feet. After leaving the saddle between East and West peaks, the route becomes unremittingly steep.

Above timberline (11,600 feet), the summit pile (Fig. 66) constitutes an additional 1000-foot elevational gain over large, unstable, angular rocks. Safe

Fig. 66: West Spanish Peak from the west ridge of East Spanish Peak. The saddle between the peaks runs along the Huerfano–Las Animas County line.

climbing calls for placing your feet with deliberate attention. A rubber-footed walking stick is very helpful. At the top of the steep approach slope, the west approach ridge levels off somewhat. The ridge is spectacular with a 6- to 20-foot-wide walkway, abruptly falling off on both sides (Fig. 67). One looks down onto the plains as from an airplane. On a clear day, one can see west to Mount Blanca, the terminal peak of the northern Sangres, and to the San Juans, 90 miles across the San Luis Valley.

On the return to the saddle, one may wish to follow the pack trail back to the Bulls Eye Mine Trail and down to the camp road; this might add 1 to 1.5 miles but may feel easier than bushwhacking downward.

A good plan for an overnight trip would be to ascend the forest trail from the parking area of the Bulls Eye Mine Trailhead and hike to the saddle between East and West peaks for an overnight camp. The following day, bag the summit and return to the trailhead. Clearly marked campsites are available.

TRAILS ON THE SPANISH PEAKS—SOUTH SIDE

CORDOVA PASS TO TIMBERLINE (AND ON TO WEST SPANISH PEAK) (SEE MAP 3)

Trailhead: Elevation 11,273 feet, 37.34833°N, 105.02444°W.

Quadrangle: Spanish Peaks 1971. 1:24,000.

Approach: At Cordova Pass, there is a trailhead leading north toward West Spanish Peak from the USFS fee parking area.

Main Theme: On the side of West Spanish Peak, a gentle 2.5-mile trail crosses broad meadows and then climbs through well-designed switchbacks to timberline on the main approach to the summit of West Peak. On emerging from the woods, one is rewarded with a spectacular view of the West Spanish Peak and valleys east, west, and south.

Description: The trail is clearly marked at the beginning and well used. About a half mile in from the trailhead is a short side trail left (to the west), called the Levy Trail. It climbs a gentle ridge and provides a view of the Sangres and the White Peaks in the vicinity of Cuchara Valley. For those who cannot walk far, this makes a nice 2-mile round-trip from the parking area with good views.

Continuing on the main trail north, this well-graded path passes through an Engelmann spruce forest and a transition to subalpine fir. The climb becomes gradually steeper. An old mine site can be visited on the left (elevation 11,298 feet, 37.35889°N, 105.01083°W). Ascend the dike ridge (west) and then descend on a short trail to a small opening.

Fig. 67: Looking west down the narrow west ridge below the summit of East Spanish Peak.

Continuing upward, one soon approaches an extensive stand of bristlecone pine where the trees become weathered, fire scarred, very large, and contorted as one climbs higher through the half-dozen switchbacks. These are perhaps the oldest bristlecones in the region among those accessible by trail. Larger ones can be seen off-trail to the east. The Apishapa Trail connects on the right and descends downhill to a trailhead on the Cordova–Aguilar Road.

Continue upward through a couple of additional switchbacks until you arrive at an abrupt timberline. Two large rock cairns mark an area where the trail ends. Climbers bent on bagging West Peak continue north with few trail markings until approaching the summit ridge.

From the cairns (elevation 11,955 feet, 37.36806°N, 105.00194°W), a great view is available in all directions; this makes a great lunch spot. Dwarf thickets of Engelmann spruce (krummholz) mark the timberline immediately to the south. In meadows here and at lower elevations, wildflower displays are nice in wet years but are not unique for the area.

WEST SPANISH PEAK (SEE MAP 3)

Trailhead, Quadrangle, and Approach: As described immediately above.

Main Theme: Continuing upward from the timberline trail, one walks ever upward (at a 35- to 40-degree angle) over the rocky south ridge, which is trailless except for occasional cairns, but the view is unobstructed. To do this, get an early start so as to reach and be off the summit by noon or soon thereafter.

Description: The one-way time to the summit is about 3 hours for a person just adequately in condition for the task, and the elevation gain is about 1600 feet from timberline. Even more than East Peak, the summit pile is quite unstable and a sturdy stick is essential for balance. All but the largest rocks move when walked on, and small avalanches are constant. Persons in the party should not be directly downhill from climbers above. On the last third of the way, as one approaches the summit, short trails and cairns appear frequently. The summit ridge is about 20 feet wide and about 100 feet long (elevation 13,563 feet, 37.37586°N, 104.99384°W). Numerous alpine flowers appear in the few accumulations of soil: larkspurs, yellow composites, penstemons, and others. The view on a clear day has to be experienced; there are no words to adequately describe it.

APISHAPA TRAIL (SEE MAPS 3 AND 16)

Trailhead: On the west side of the road between Cordova Pass and Aguilar, having descended about halfway to the plains.

Quadrangle: Cuchara Pass, Spanish Peaks 1:24,000.

Approach: From Cordova Pass (mile 0.0), continue east and downslope toward Aguilar. Go through the tunnel in the dike (mile 3.7), and reach the marked trailhead on the left side of the road (mile 5). Continue around the bend an additional 0.1 mile to the parking area on the right. Walk uphill from the parking area through the trees on an obscure path to gain the road beyond the bend, and you will recover the actual trailhead of Trail 1324 (elevation 9710 feet, 37.34354°N, 104.99301°W).

Main Theme: In terms of covering most of the vegetation zones in the area, this is one of the most satisfying trails. There are dense woods, open woods, and streams, especially at lower elevations. Higher up, great views open up in various directions. Wildflower displays are good. At the upper end, the trail connects to the West Peak timberline trail described above. The total distance, not marked, is about 3.5 miles one way, with an elevation gain of nearly 1500 feet.

Description: The trail is well designed, of gentle grade, and well maintained. The lower half of the trail is through mature or recovering forest. All the usual suspects are here: Douglas firs, limber pine, bristlecone pine, a few white firs at lowest elevation, common juniper, aspens, subalpine fir, and a few ponderosa pines. Flowers include blue columbine, red penstemon, sedum, and red paintbrush.

About 1.5 miles up the grade is a sign, "Wahatoya TH. Built by Volunteers for Outdoor Colorado, US Forest Service." That trail goes east through the heart of the wilderness to the saddle between East and West peaks and connects with the Wahatoya community on the north side. Shortly after the sign, the trail opens up to frequent views east and south. Continuing west, the trail crosses a north-south–oriented dike and proceeds parallel to it to the north. The trail emerges onto an open area with several acres of rock and meadow. The view of West Peak to the north is excellent (37.35810°N, 104.99893°W, elevation 11,193 feet). This is a good lunch stop.

The trail then bends west and joins the West Peak approach trail within a half mile (37.36100°N, 105.00646°W, elevation 11,340 feet). In this vicinity, a sawn tree trunk, 10 inches in diameter, had about 120 growth rings.

SHAFER TRAIL #1391 (SEE MAP 3)

Trailhead: Elevation 10,947 feet, 37.34917°N, 105.04250°W.

Quadrangle: Cuchara Pass 1967, 1979. 1:24,000.

Approach: From Cuchara Pass, proceed up the Cordova (Apishapa) Pass Road 4.8 miles to a trailhead on the left (north) side of the road. The wide spot on a curve provides parking for perhaps two (or three small) cars. The trail has an information board with a map and sign-in book.

Main Theme: This trail crosses the west side of the Spanish Peaks wilderness, heading west and gradually north. One hikes down through mature conifer forest that seldom changes.

Description: Some maps may call this the "Chaparral TH," but the roadside sign calls it the "Shafer Trailhead." The road sign notes that the trail is blocked 3.5 miles in by private property. This trail appears to be the only way to approach South, Middle, and North White Peaks, which are U.S. National Forest land but where access is blocked from most other directions. The White Peaks can only be approached by compass and bushwhacking by experienced hikers. Take the Shafer Trail 1.5 to 2 miles in, then bushwhack west to reach the White Peaks.

The Shafer Trail is a well-planned, mostly gentle downward walk (except when returning), beginning through a mature Engelmann spruce forest with low bush blueberry understory, some golden banner, and little else. After 15 minutes, you come to a gap in an impressive dike where narrow photogenic views of West Peak are framed. The forest changes slowly, and you pick up mixed-age trees of Douglas fir, subalpine fir, bristlecone pine, and mature aspen. Limber pines occur on exposed rocky places, as well as occasional lodgepole pines. It is interesting that lodgepoles occur in mixed forests at this southern extreme of their range.

In general, the woods have the aspect of a mature forest where fire has not been present for perhaps 120 years. On and on it goes without view or respite. It's not a good choice if views and variety are your thing, but it's fine if you are looking for some quiet time in the woods. Judging by the logbook, this trail is little traveled.

NORTH FORK OF THE PURGATOIRE—SOUTH END (SEE MAP 2)

Trailhead: 9820-foot elevation, 37.255°N, 105.10944°W.

Quadrangle: Stonewall 1979. 1:24,000.

Approach: Taking Route 12 from La Veta, cross Cuchara Pass and proceed 8.5 miles, past North Lake and Monument Lake, to the turnoff on the right (north). The gravel forest road North Fork Road (or FR 411) proceeds 4.2 miles to the well-marked trailhead. The trail is TR 1309 on the atlas.

Main Theme: A trail with a gentle grade that generally follows the stream-sized North Fork of the Purgatoire. It passes through woodland and a succession of meadows with great wildflower displays all summer. About 560 feet of elevation gain as described here.

Description: The trail is generally well graded although rocky in places. Horse traffic is light. At the trailhead a sign reports, "Trinchera Peak road 5 miles, Blue Lake campground 6.5 miles." As described here, it is a nice half-day walk, about 2 to 3 miles one way. It winds through patches of older Engelmann spruces, and through sparse aspen parks with thin canopy and lush meadow beneath. Visibility extends to hundreds of feet in places. After the first forested section, the trail winds through three successive meadows, each larger than the last (Fig. 68). There are three stream crossings for which a walking stick is helpful in navigating pole crossings or stepping stones. Several local willow species define the edges of streams and meadows.

Flowers in the lower reaches are fireweed, elderberry, arnica, and cow parsnip. Higher up are veratrum, monkshood, buttercup, clover, geranium, white mallow, white daisy, several yellow daisies, Indian paintbrush, harebell, and Indian soldiers. Near the last creek crossing is elephant head *Pedicularis*, not frequently found in this region. At the end of the third meadow, about 15 to 20

Fig. 68 (left): The second meadow along the North Fork Trail.

Fig. 69 (below): North Fork Trail, showing a portion of the large beaver dam complex that covers many acres.

acres (37.27305°N, 105.12194°W, elevation 10,380 feet), there is an especially lush display of many blooming species including cinquefoil and mariposa lily.

About midway, on the right side of the trail and partially hidden from view by a brushy border, an unusual feature is one of the largest beaver dam complexes you are likely to encounter (Fig. 69). The dam is more than 8 feet high on the lead edge and consists of a few hundred feet of branch dams. Obviously the work of generations of beavers, the family's patriarch must have been trained by the U.S. Army Corps of Engineers!

WESTERN HUERFANO COUNTY

MOSCA PASS TRAIL (SEE MAP 10)

Trailhead: An ample parking area on the west side of the pass itself (see Chapter 7, Drives, Drive #4B, Mosca Pass).

Quadrangles: Mosca Pass, Medano Pass 1:24,000.

Approach: From the top (east end), use the Mosca Pass Road described in Chapter 7, Drives. On the west side, the bottom of the trail, there is a trailhead that begins the eastward climb from near the Great Sand Dunes National Park and Preserve Visitor Center.

Main Theme: Begun at the top, the trail is well planned and maintained. The downward slope is gentle in the upper portion, getting steeper toward the Great Sand Dunes. Various ecological communities are traversed. It features nice views and interesting geology.

Description: Just below Mosca Pass (elevation 9175 feet), the trail into the nature preserve begins on the west end of the parking area, 100 yards west of the pass itself. The forest is open parkland with aspen, ponderosa pine, and limber pine populations of various ages. Some of the largest and oldest limber pines to be seen in the region occur as one follows the trail downward.

From the top, the sandy trail descends gently through wet areas with willow thickets and dry, open areas. As one approaches Great Sand Dunes National Park, the steeper trail develops wider views of the San Luis Valley and the dunes. Here and there, try climbing the rocky north side of the canyon for improved photogenic views, but be careful to steer clear of cacti in your search for the good photo (I usually back into them while focusing) (Fig. 70).

LILY LAKE/UPPER HUERFANO BASIN (SEE MAP 17)

Trailhead: Elevation 10,000 feet, 37.62360°N, 105.47295°W.

Quadrangle: Blanca Peak 1:24,000.

Fig. 70: Upper portion of the Mosca Pass Trail, looking toward the sand dunes. This is a good example of a mixed conifer and aspen woodland.

Approach: (See also Chapter 7, Drives, Drive #4C, Upper Huerfano Basin.) From Gardner (Kay's Country Store), proceed west out of town on Route 96. Beyond the edge of town (0.7 miles), turn left on the road to Red Wing. Ignore the left turn to Malachite (Pass Creek Road, mile 5.2), and continue westward. Later, another road branches left, but continue to bear right. A Mosca Pass sign occurs as does an FR 480 sign obscurely on the right, a block beyond the intersection. One mile later, pass the old log and adobe ruins of "Archuletaville" on the right. Finally arrive at the Y-intersection that is Sharpsdale. This "town" is now only a well-kept one-car garage with a flat roof. A sign says, "Mosca Pass, right, 6 miles." To the left is Lily Lake, 10 miles, and the forest boundary (upper Huerfano), 8.5 miles. Turn left. The road is 580 southwest to FR 408, to TR 1308.

The next 10 miles to the trailhead for Lily Lake will take more than an hour to traverse. See Chapter 7, Drive #4C for details of this route. 4WD is not required, but high ground clearance certainly is, and a limited-slip differential is a plus. About 5 miles in, the views open up and subalpine meadows allow good vistas. The trailhead has ample parking space.

Main Theme: The hike moves through the most ruggedly scenic alpine valley in this vicinity, or anywhere else in the Rocky Mountains. It is a large J-shaped cirque basin surrounded by a cluster of four of Colorado's fourteeners: Blanca Peak, Ellingwood Point, Mount Lindsey, and Little Bear Peak, plus the Iron Nipple, and a number of imposing, slightly smaller features along the J-shaped ridge. Mountain sheep and deer are seen in the valley.

Description: The round-trip to Lily Lake from the trailhead is probably about 10 miles, although no distances are offered on signs and the elevation gain is nearly 3000 feet. Almost immediately after leaving the trailhead, the trail opens out into a large meadow. It drops down off the terminal moraine into the flat Upper Huerfano River Valley and continues for about 1 mile to a fork. The Upper Huerfano Trail continues straight for 1 mile to an old mine site with heavy equipment and a stack of firewood laid up for a hopeful future. The other (right) fork is marked "Lily Lake" and points to the west flank of the valley. Although steep in spots, the trail proceeds south and gradually angles up the west slope. Views are mostly open for the next couple of miles with 4000-foot ramparts and nearly sheer cliffs to be seen to the east and south. An odd observation is that the conifer trees are almost exclusively Engelmann spruce. The lack of tree diversity is probably explained by the fact that 19th-century mining involved extensive lumbering for firewood and mine timbers. However, the recovered spruce forest is now of high quality and in excellent health. The last mile of the trail is steep, involving several switchbacks that climb west, perhaps 1000 feet. Lily Lake is at the top of a cirque basin and is an impressively large alpine lake surrounded by rock and a krummholz zone (elevation 12,900 feet, 37.5948°N, 105.4901°W).

Timing: In our late June hike, the 350-inch snow accumulation of the previous winter left 4 to 6 feet of snow pack on the upper trail and the lake remained 90% frozen. After early July, you will find the best wildflower displays and be sure of finding the upper trail. However, regardless of whether you make the lake your objective, this basin is an outstanding walk. Finally, it is best to get an early start considering the slow drive approaching the trailhead. Being on the trail by 9:00 AM would be best so as to be out of the high country by noon, thereby avoiding the danger of thunderstorms.

NORTHERN HUERFANO COUNTY

GREENHORN TRAILS (SEE MAPS 6 AND 7)

Trailhead: A parking lot, as described in Chapter 7, Drives, Drive #3 approach.

Quadrangles: San Isabel, Badito Cone 1:24,000.

Main Theme: These trails are on the slopes of the Wet Mountain Ridge, which is dominated by Greenhorn Peak. The ones discussed here are at timberline and above. The ancient geology, wildflower meadows, and old timber in the region are inspiring.

Description: Two trails lead into the wilderness area (Fig. 71). At an information sign, trailhead 1316 proceeds toward the summit, on a compass bearing of about 80 degrees. The elevation gain with steady upward climb is nearly 900 feet. Views over the adjacent valley are wonderful all the way up. Gardner and other settlements are easily seen on clear days.

The second trail is less physically ambitious but very rewarding. It is unlabeled and heads east on a bearing of 120 degrees from the parking lot. Wider than the usual trail, it is apparently an old service road that follows a contour below the summit at the level of the parking lot. It meanders through the wilderness area and a mature spruce forest that seems to be old growth in places. Wildflower displays are excellent. Nodding yellow-flowered thistles (*Cirsium scopulorum*) (Fig. 72) are common here, as are penstemon, harebell, yellow members of the aster family, cream-colored paintbrush, cinquefoil, iris, wallflower, yarrow, alpine clovers, asters, fleabane daisies, and numerous butterflies attending the pollination duties.

Fig. 71: View of Greenhorn Peak from the trailhead parking lot, where Jody Keating inspects the sign. The wilderness begins just to the right side of this view.

About 1 mile along, one arrives at a large fine meadow with picturesque weathered rock outcrops, photogenic forested gullies (Fig. 73), and distant views. Not unlike the surface of Pike's Peak, much of the mountain seems to be built of ancient quartz and pink feldspar, often with black streaks and large crystals. Timberline on this side is abrupt.

TO THE EAST

DINOSAUR TRACK SITE

Trailhead: On the north end of the trail, leave from the Withers Canyon Access site, south of La Junta, Otero County.

Quadrangles: Riley Canyon, Higbee 1:24,000.

Approach: See Chapter 7, Drives, Drive #3, Greenhorn Peak and Wet Mountains Loop. Good gravel roads end at a timberline parking lot. The road gets a little rougher toward the end but is still passable, carefully, by car. The elevation of the parking lot is 11,456 feet.

Main Theme: This attraction is peripheral to this guide's territory, but for the adventuresome it provides a nice contrast to alpine scenery and is worth the extra miles. The track site, along both sides of the Purgatoire River, is presented as the largest collection of dinosaur tracks in the world. As described here, it is reached at the end of a 5.3-mile (10.6 miles round-trip), mostly level walk through the Purgatoire Canyon, a broad, low-elevation, arid landscape. The track site is part of the 16,000-acre Picketwire Canyonlands tract. In 1991, it was transferred to the Comanche National Grassland administered by the USFS.

Before your trip, it is strongly advisable to check with the Comanche National Grassland agency office in La Junta for advice on current conditions, and arrangements for guided 4WD tours if desired (see Resources). The site is open from dawn to dusk, and there is no camping in the canyon. The temperature is more hospitable in spring and fall.

The Purgatoire Canyon is broad but not deep, a gentle landscape with an entrenched river surrounded by rolling bottomlands. Boulders stand on the adjacent rocky slopes that rolled down from the rim rock defining the canyon. Picturesque old gnarled cottonwoods are common, as are cacti. Wildflower displays are good in spots along the trail and on the adjacent slopes. Although the bottomlands now appear as healthy grassy meadows, they are former rangelands where a century of persistent overgrazing destroyed numerous native species, leaving a dominance of Eurasian weeds and grasses.

Fig. 72 (left): The yellow-flowered nodding thistle (*Cirsium scopulorum*) on the side of Greenhorn Peak.

Fig. 73 (below): Branch of very old Engelmann spruce, bearing a heavy load of cones.

Approach: From Walsenburg, take Route 10 to La Junta, about 70 miles. From La Junta, take Highway 109 south for 13 miles; right on County 802 for 8 miles; left (south) on County 25 for 6 miles; left on Forest Service Road 500A for ¾ mile. Proceed through a wire gate, and take the left fork to the parking area and bulletin board. This is the Withers Canyon access site. On the approach, early in the day and at dusk, jackrabbits are common, as are families of pronghorn antelope.

Get going early as there is no camping on site. Good signage marks the way to the trailhead and occurs within the canyon itself. A U.S. Geological Survey quadrangle or an atlas will be very helpful. The last portion of the trip is best approached using a high-ground–clearance vehicle. Hike from the locked pipe gate, descending 500 feet, after which the hiking is mostly level. Follow "To track site" signs.

Description: There are two ways to see this site. For a fee, guided tours are offered by Forest Service personnel on Saturdays during the summer season. On these tours, individuals or families furnish their own vehicles and gain access to the site through Pinon Canyon Military Reservation at the south end. The tours visit the abandoned Rourke ranch site and then drive north along the Purgatoire River to the dinosaur track site.

As described here, the track site can be reached year round from the Withers Canyon Access site as a 10.5-mile round-trip hike. Along the way, you will see ruins of 19th-century homesteads with partially standing stone or adobe walls. There are walls of the picturesque Delores Mission and Cemetery, built between 1871 and 1889 by Mexican pioneers. A few headstones can still be read. In various locations are fine examples of petroglyph rock art by Native American artists between 1000 and 1800 AD. They occur on large stone blocks or along the canyon walls. Because they are fragile, their locations, by policy, are not published. You will need to explore off trail in likely sites. The rock art is found on flat, slightly overhanging rock faces.

From the parking loop, the trail descends steeply into Withers Canyon, a tributary of Purgatoire Canyon. After reaching the canyon floor, the trail is mostly level. After about 1.5 miles, the river comes into view and the trail follows it occasionally (Figs. 5 and 6). Upon reaching the track site, there is a parking area and comfort station used by the driving tours from the south. Descriptive, illustrated signs are located at the head of a short side trail that brings you to the river and the dinosaur tracks.

The giant footprints are large depressions in Jurassic limestone (Morrison Formation) (see Schumacher, 2004) formed about 150 million years ago, and were originally laid down on shoreline mud by allosaurs and brontosaurs. Altogether, about 1300 tracks represent several species in about 100 different trackways. In one case, the tracks show individuals traveling together in groups (gregarious behavior), the only known location documenting herding behavior.

The river divides the track site into east and west portions. Perhaps the largest display is reached by wading across the river. To do so, bring shoes or sandals suitable for wading across a rocky bottom. Crossing is definitely not safe during high water such as during the spring runoff. Check first with the Comanche Grasslands office in La Junta regarding river conditions (see Resources).

Recommendations: One must be in good hiking condition, as the final approach must be walked. A good plan is to stay overnight in La Junta and then leave early for the all-day walk in the canyon.

This trail is best taken some other time than summer. The author's early June timing was not optimal that year. Although cool enough, the river was swollen, making it unsafe to cross. Many dinosaur tracks on the near side were covered with mud from recent spring flooding. The fall season should be best, as rains will have cleaned the tracks and the river can be safely waded. Not seeing lots of tracks does not spoil the trip, since the walk through the canyon is pleasant and photogenic. It displays a notably different ecology from areas above 6000 feet.

Preparations: This can be a demanding hike when the temperature gets into the 80s and above—this is not an alpine environment. The U.S. Forest Service recommends carrying 1 gallon of water per person. If you drink less, you risk serious dehydration, especially because of possible summer temperatures of 95°F to 105°F. You should be in good physical condition and acclimated to heat if you try this walk in midsummer. Carry lunch, sun protection, a parka, and a first aid kit.

At this high plains (low) elevation, badgers and rattlesnakes are present, as are tarantulas and scorpions, although we saw none. When you stop to rest beneath trees, flies or other insects may be nuisances. If you venture off the trail, carry a walking stick to stir the high grass ahead of you, thereby announcing your presence to any snakes that might be there. Pay attention to where you sit or place your hands when exploring. No camping is allowed in the canyon, although we saw pickup campers at the trailhead entering Withers Canyon.

Maps, brochures, and information on current conditions are available at the Comanche National Grassland headquarters in La Junta (see Resources).

CHECKLISTS OF ORGANISMS OF THE SPANISH PEAKS REGION

CHECKLISTS are valuable for narrowing the choices when used with regional floras and faunas, especially when used by persons with some background and experience. In some cases, a few features and ecological preferences are given that provide clues. Among the lists that follow, the most detail is offered for trees, because they define the ecology of this landscape. Everyone should aspire to knowing the trees and other common woody plants in this area, a comparatively short list.

The lists of organisms that follow vary in accuracy, as they are based on different published sources. They should be viewed as provisional because additional fieldwork in this area has the potential to modify these lists considerably. Usually more than one source has been cross-checked with available literature, as well as with the author's experience.

Organisms that are found in Huerfano County are marked [H]; those found in Las Animas County are marked [L]; those found in both counties are marked [HL].

CHAPTER 9

FERNS AND OTHER SPORE-BEARING VASCULAR PLANTS

THIS LIST is compiled from Kartesz and Meacham, *Synthesis of the North American Flora* (1999), and from Flora of North America Editorial Committee, *Flora of North America* (1993: volume 2).

SPIKE MOSSES, FAMILY SELAGINELLACEAE

Dense spike moss
Selaginella densa Rydb. [L]

Blunt-leaf spike moss
Selaginella mutica D. C. Eaton ex Underw. var. *mutica* [HL]

Underwood's spike moss
Selaginella underwoodii Hieron. [HL]

Weatherby's spike moss
Selaginella weatherbiana R. M. Tryon [H, rare]

SCOURING RUSHES, HORSETAILS, FAMILY EQUISETACEAE

Field horsetail
Equisetum arvense L. [H]

Common scouring rush
Equisetum hyemale L. var. *affine* (Engelm.) A. A. Eaton [H]

Smooth scouring rush
Equisetum laevigatum A. Braun [HL]

ADDER'S TONGUES, FAMILY OPHIOGLOSSACEAE

Echo moonwort
Botrychium echo W. H. Wagner [H, rare]

Western moonwort
Botrychium hesperium (Maxon & R. T. Clausen) W. H. Wagner & Lellinger [H, rare]

Pale moonwort
Botrychium pallidum W. H. Wagner [H, rare]

MAIDENHAIR FERNS, FAMILY PTERIDACEAE

Southern maidenhair
Adiantum capillus-veneris L. [L, rare]

False cloak fern
Argyrochosma [Pellaea] fendleri (Kunze) Windham [HL]

Southwestern cloak fern
Astrolepis integerrima (Hook.) D. M. Benham & Windham [L]

Eaton's lip fern
Cheilanthes eatonii Baker [L, rare]

Slender lip fern
Cheilanthes feei T. Moore [L]

Fendler's lip fern
Cheilanthes fendleri Hook. [L]

Beaded lip fern
Cheilanthes wootonii Maxon [L, rare]

Star cloak fern
Notholaena standleyi Maxon [L, rare]

Purple cliffbrake
Pellaea atropurpurea (L.) Link [L, rare]

Smooth cliffbrake
Pellaea glabella Mett. ex Kuhn subsp. *simplex* (Butters) Á. Löve & D. Löve [L, rare]

Wright's cliffbrake
Pellaea wrightiana Hook. [L, rare]

BRACKEN FERNS, FAMILY DENNSTAEDTIACEAE

Western bracken
Pteridium aquilinum (L.) Kuhn var. *pubescens* Underw. [HL]

SPLEENWORTS, FAMILY ASPLENIACEAE

Ebony spleenwort
Asplenium platyneuron (L.) Britton, Sterns & Poggenb. [L, rare]

Black-stem spleenwort
Asplenium resiliens Kunze [L, rare]

Forked spleenwort
Asplenium septentrionale (L.) Hoffm. rare [HL]

Spleenwort
Asplenium trichomanes L. subsp. *trichomanes* [L]

WOOD FERNS, FAMILY DRYOPTERIDACEAE

Fragile fern
Cystopteris fragilis (L.) Bernh. [HL]

Southwestern brittle fern
Cystopteris reevesiana Lellinger [HL]

Male fern
Dryopteris filix-mas (L.) Schott [L]

Northern holly fern
Polystichum lonchitis (L.) Roth [H]

New Mexico cliff fern
Woodsia neomexicana Windham [L, rare]

Woods fern
Woodsia oregana D. C. Eaton subsp. *cathcartiana* (B. L. Rob.) Windham [HL]

Plummer's cliff fern
Woodsia plummerae Lemmon [L, rare]

POLYPODYS, FAMILY POLYPODIACEAE

Rocky Mountain polypody
Polypodium saximontanum Windham [H]

WATER CLOVERS, FAMILY MARSILEACEAE

Hairy water clover
Marsilea vestita Hook. & Grev. [L]

CHAPTER 10

GYMNOSPERMS

CONIFERS

JUNIPERS, FAMILY CUPRESSACEAE: GENUS *JUNIPERUS* L.

Junipers have small, juicy, berry-like cones that may appear green or blue, often having a waxy bloom. Leaves are often of two forms. Scale-like leaves are less than ⅛ inch and are pressed, shingle-like, against each other. On branch tips, juvenile leaves may occur that are pointed and needle-like, up to ½ inch long but usually shorter than ¼ inch.

Common juniper

Juniperus communis L. var. *depressa* Pursh

Key Features: A spreading shrub up to 1 yard high, having only the elongate (ca. ½ inch) needle-like leaf form. Leaves have a white line on the hollow side (Fig. 74).

Range and Ecology: A widespread species with populations throughout the boreal (northern) forest up to the tundra tree line in Canada, ranging from Nova Scotia to Alaska, to the Pacific Northwest. Scattered populations occur in the central Rockies, as well as throughout the Front Range and down the entire Sangre de Cristos and Spanish Peaks. The species is documented from Huerfano County. There are only a few small populations in New Mexico and Arizona and farther south.

These shrubs often form wide patches, especially under scattered conifer trees, over a wide elevational range up to timberline.

Fig. 74: Common juniper (*Juniperus communis*) clump along the Spring Creek Trail.

One-seeded juniper

Juniperus monosperma (Engelm.) Sarg.

Key Features: Branches with sharp (juvenile) leaves at branch tips and tiny scale-like leaves lower on the branches. Always found among pinyon pines above the "lower timberline."

Range and Ecology: The species is more frequently found in Arizona and New Mexico and has its northern range limit in central Colorado. It also reaches the Texas and Oklahoma panhandles. It is found in better-watered sites of lower to mid elevations, and it constitutes the lower or dry timberline. It is the juniper of this area's pinyon-juniper (P-J) ecosystem and is documented in both counties.

Trees are often picturesquely contorted, growing as high as 6 feet or more, but usually are shorter and often shrub-like with multiple trunks from the base. Five-hundred-year-old trees are not unusual. The Methuselah tree at Garden of the Gods is estimated to be about 900 years old and has several huge trunks.

This juniper has much local value. The large, calorie-rich "berries" are favored by wildlife such as chipmunks, quail, foxes, squirrels, and deer that compete with local people who also eat them, when ripe, in the winter. Navajo weavers used the bark and berries to dye wool green. Wood of the species is favored for use in cooking fires, and it produces fragrant smoke. In the old days, its wood constituted the main fuel supply of Taos and Santa Fe.

American Indians used the wood to make war sticks, prayer sticks, war bows, papoose cradle canopies, and charcoal for smelting jewelry. The wood, divided into reddish heartwood and white sapwood, is close ringed, hard and dense, and resistant to decay when dry. As with other "cedars," the wood is fragrant. It is used for fence posts locally but is seldom straight enough for commercial exploitation.

Rocky Mountain juniper

Juniperus scopulorum Sarg.

Key Features: Similar to the above species but with large-scale leaves. It is less shrubby in appearance, with a straighter dominant trunk, and ranges beyond the P-J formation.

Range and Ecology: The range extends from British Columbia throughout the Rockies into New Mexico, Arizona, and the Texas Panhandle. It grows throughout the Sangre de Cristo Range and Spanish Peaks, in both counties, and extensively throughout lower elevations of 5000 to 8000 feet. It occurs extensively throughout higher elevations in central and western mountainous Colorado. On mountain slopes, the trees favor canyon bottoms, dry exposed mesas and cliff talus slopes, and the long slopes of the Great Plains–Rocky Mountain transition.

The species is a close relative of the eastern red cedar (*Juniperus virginiana* L.) and was not originally distinguished from it. When open grown, it forms a narrow 10- to 20-foot-tall tree with a narrow, rounded crown. In sheltered canyons, it may reach 30 feet high and have a weeping form. Bark is weathered gray and twisted and reddish brown beneath. Twigs are four angled with leaves in pairs, alternating with adjacent pairs at right angles. Scaly leaves are about ⅛ inch long and dark or gray-green. The berries, clear blue under a thin-skinned, whitish, waxy coat, are 0.25 to 0.33 inch in diameter.

The red and white wood is very handsome and fragrant like its close eastern relative. The trees are not big enough or abundant enough to be commercially exploited.

Questionable record: Kartesz and Meacham (1999) record *Juniperus virginiana* var. *virginiana* as occurring in Las Animas County. This would be an isolated occurrence, outside the range that ends mostly in the central Great Plains.

CONIFERS, FAMILY PINACEAE: GENERA *ABIES* MILL., *PICEA* A. DIETR., *PINUS* L., *PSEUDOTSUGA* CARRIÈRE

In our region, all species of this family are trees. Leaves are needle shaped and evergreen, and normally persist for several years. Seeds are produced in female cones that have woody scales. After some acquaintance, these species usually can be identified easily from some distance.

FIRS, GENUS *ABIES* (TWO SPECIES)

Leaves are flattened needles arising singly from twigs. The needles have a green base at the point of attachment. Cones stand up on the branches and usually disintegrate after seeds are dispersed, or when dried.

White fir

Abies concolor (Gordon & Glend.) Hilldebr. var. *concolor*

Key Features: Needles are largest of our fir species, more than 1.5 inches long (Fig. 75). They are broad and silver-green, especially when young, and tend to turn upward, especially near the branch tips.

Fig. 75: White fir branch tips.

Range and Ecology: Populations are scattered at higher elevations through the Sierras, in central Utah, and less frequently in Nevada. In Colorado, they occur extensively in the San Juans, the entire Sangre de Cristo Range, and the Wet Mountains. Trees are scattered at higher elevations. Ours are the easternmost populations. They are documented from Huerfano County.

A large tree with symmetrical and broad conical shape, the species is very ornamental when open grown. It grows well in drier mountain habitats in areas of deep snow and winter storms. The tree is quite fragrant and a favorite of animals. Seeds are eaten by squirrels and grouse, porcupines like the bark (Fig. 76), and deer will browse the leafy branches. The species is cut for construction and classed as "fir." Economically, the species makes a minor contribution due to its scattered distribution.

Rocky Mountain subalpine fir

Abies bifolia A. Murray (= *A. lasiocarpa* (Hook.) Nutt.)

Key Features: The trunk is mast straight and grows up to 75 feet high or much shorter near timberline. Whether open grown or confined to a forest, the crown is always narrow and very slim tapered above. Leaves are 1 to 1.75 inches long. Seed cones, 2.25 to 4 inches long, are densely clustered with broad

Fig. 76: Close-up view of white fir bark. The tree is probably more than 100 years old.

purple scales. Older bark is thin, pale light gray, and has narrow, brownish fissures. Twigs and needles are both stiff rather than flexible.

Range and Ecology: An extensive western range. They occur northward through Yukon Territory, with populations in southern Alaska, southward through the Rocky Mountains. Populations are found at higher elevations in the entire Sangre de Cristo and Wet Mountain ranges, ours being among the most southerly. While not documented, they certainly occur here, mostly above 11,000 feet, mixed with Engelmann spruce. They reach timberline in some areas.

Subalpine firs can also grow in windless, dense groves in protected sites. The trees are seldom cut for lumber, and the weak wood has negligible economic value.

SPRUCES, GENUS *PICEA* (TWO SPECIES)

Spruces have four-angled, pointed needles attached singly to twigs. At the point of attachment, mature leaves are always constricted and brown. When a twig specimen is allowed to dry, all needles fall off. Cones have thin scales and are pendant (hang down from) branches. When dried, cones do not fall apart.

Engelmann spruce

Picea engelmannii Parry ex Engelm. var. *engelmannii*

Key Features: Cones are up to 2.5 inches long (6 cm) and are flexible with jagged, toothed scales. Young twigs are lightly and finely hairy. Bark is only ¼ to ½ inch thick. Leaves are flexible, up to 1⅜ inches long, and blue-green, sometimes without resin ducts. Tips are not prickly to touch (Fig. 77).

Range and Ecology: Widespread in the northern Rockies into British Columbia and the Cascade Range at lower elevations. This spruce extends from about 8000 feet up to timberline in mountainous Colorado and down the Sangre de Cristo Range. While not locally documented, the species occurs in the Wet Mountains and Spanish Peaks. Scattered populations occur in New Mexico and Arizona.

As one of the large and dramatic tree species of the Rockies, they form dense, fragrant, sometimes pure stands in forests and reach as high as 100 feet. The tree always appears dark or silhouetted at sunset. They regularly reach timberline and survive as dwarfed, sprawling trees. Relentless winds prune the windward side, which, combined with leeward living branches, gives them a flag-form appearance. The rooting capacity of branches in contact with the

Fig. 77: A branch ending of Engelmann spruce in good condition.

ground causes the trees to move across the tundra in some places in the central Front Range. This phenomenon was recorded on Niwot Ridge by the late University of Colorado botanist, John Marr.

The seedlings can endure years of growth in deep shade, a characteristic of a climax forest species. When light appears in the canopy, the seedlings grow very quickly toward the opening.

The strong and light wood is harvested and favored for use as utility poles. Unfortunately, when ignited in the forest, their pitch-laden wood burns hot and fast, quickly going out of control.

Colorado blue spruce

Picea pungens Engelm.

Key Features: Young twigs smooth. Leaves extend in all directions and are sharp pointed and sometimes but not always light blue. Seed cones are longer than 2.5 inches (6 cm). The bark can get up to 15 inches thick and becomes deeply furrowed. Twigs are stout and orange- to gray-brown. Leaves extend in all directions, are prickly with long tips, and up to 1.25 inches. Their signature light blue-green or silvery color is due to a waxy (glaucous) cuticle, and the effect varies genetically among individuals. Also, with time, needle appearance weathers to a darker green.

Range and Ecology: One of the continent's most restricted tree species, it is nowhere common. Scattered populations occur in eastern Idaho, western and southern Wyoming, Utah, Arizona, and New Mexico. Most occur in the Colorado Rockies including the Wet Mountains and Sangre de Cristos and Spanish Peaks (although not officially documented). In the north, they occur up to 9000 feet; in the south, up to 11,000 feet. Populations are almost certainly rare in Huerfano County. I have found only the occasional tree.

Blue spruce is very handsome and of classic conifer proportions when open grown. In forest competition for light, the pole-like trunk carries an irregular crown pruned by the elements. Throughout the country, blue spruces are probably the most planted ornamentals among western conifers. In Moscow, Russia, the species has been planted widely as an ornamental. The wood is soft, brittle, and full of knots, and therefore not often harvested.

PINES, GENUS *PINUS* (FIVE SPECIES)

Pine trees have clusters of two to five needles bound together in fascicles that arise from short lateral shoots. The needles are always longer than those of spruces or firs. Cone scales have expanded woody tips.

Bristlecone pine, foxtail pine

Pinus aristata Engelm.

Key Features: Needles in fascicles of five, bearing scattered waxy white spots. Branch ends have closely spaced needles extending in all directions like a bottle brush. Young mature cones are purple and usually occur in pairs near branch ends. Mature cones, up to 3.5 inches, have long, curved bristles that emerge from the center from each cone scale. The bark is brown and fine textured (Fig. 78).

Range and Ecology: Once considered a single species ranging from the Colorado Rockies to the White Mountains of California, these long-lived pines are now divided into two species. The above Latin name is restricted to the Colorado populations. They are considered as among the rarest trees on the continent. Occurring in the Spanish Peaks and the crest of the Sangre de

Fig. 78: Bristlecone pine (*Pinus aristata*) branch with immature second-year cone bearing scales with bristles.

Cristos, they are documented only for Huerfano County but also occur in Las Animas County. They occur as far north as Colorado's Front Range.

Natural populations are found at upper timberline on the highest and most exposed sites, enduring exposure to severe weather conditions. They can survive a 2- to 3-month growing season and a mean annual temperature of 2°F to 3°F above freezing. Locally, they occur in scattered stands in the Sangre de Cristos and the Spanish Peaks, usually beginning above 11,000 feet.

In many timberline localities, this species forms stunted, gnarled trees, often called krummholz or wind timber. Specimens are especially noted for their great age, which was shown to exceed 2400 years at Windy Ridge in Guffey County, Colorado. In one 8-inch-long core sample I took from a recently dead tree in central Colorado, I counted 700 growth rings; it began growing on that site about 1300 AD! However, most local populations I have seen are "young," or less than 200 years old. In an off-trail location on West Spanish Peak, I have seen specimens that may be 600 years old or more.

When planted in protected localities at lower elevations, bristlecones form a luxuriant, handsome conical crown. In their usual extreme habitats, their growth is highly irregular and bent away from the prevailing wind. Wind-blasting of the trunk frequently exposes extensive areas of weathered wood. On the lee side, narrow strips of living bark may be all there is to support the nutrient requirements of the remaining crown. The winged seeds are carried by strong winds to distant peaks.

Native stands are seen by few people because they occur above most roads. When encountered by hikers, such as on the main approach trail to West Spanish Peak, they are universally regarded as very picturesque.

Lodgepole pine

Pinus contorta Douglas ex Loudon var. *latifolia* Engelm.

Key Features: Needles are two to three per fascicle and shorter than 3 inches (8 cm). Cones are strongly bent or contorted.

Range and Ecology: Found most extensively in the Canadian Rockies, to the south they divide into a variety found in California's Sierras, as well as an eastern variety found in Wyoming, Utah, and Colorado. In northwest Wyoming, in Grand Teton and Yellowstone national parks, they form extensive stands that cover hundreds of square miles. The southern range limit is in the Sangre de Cristos of Huerfano County, in the Spanish Peaks, and on the Las Animas side of Cuchara Pass, despite the fact that they are not documented this far south. Here, at their southern range limit, populations are smaller and more scattered.

Lodgepole pines form the classic "fire forest." Scales on their contorted or twisted seed cones are sealed by pitch and retained on the branch for many years. Heat is required to melt the pitch and thus release the seeds. Therefore, wherever stands occur, lodgepoles arose from a charred landscape with all individuals being about the same age. After a burn, as many as 100,000 seedlings per acre become established, and survivors endure stiff competition for space and light.

Like aspens, they are called "pioneer trees." In the absence of fire for, say, a century, they are eventually replaced by spruces or firs that invade the understory. These replacement species are shade tolerant and grow slowly and persistently. In our area, large populations occur south of Raspberry Mountain, as well as on top of Old La Veta Pass. A few smaller populations are encountered throughout the Spanish Peaks at lower elevations.

In the north, lodgepoles characteristically form tight stands of spindly trunks; walking among them can be nearly impossible due to their interlocking dead branches. In Wyoming, locals call them "dog hair" pine since they grow "thicker than hair on a dog's back." Dead trees can't fall to the ground but instead lean against neighbors, eventually forming a jackstraw pile. Stands older than 100 to 150 years are rare. In this area, their stands are more open than those farther north.

The straight young trees were favored by the Plains tribes as portable lodge poles, hence the common name. Small pines were cut to yield poles about 15 feet long by 2 inches in diameter at the butt. When debarked and seasoned, they formed straight, uniform, light tepee poles that were easily dragged by travois from camp to camp. Later settlers used them as fence posts, cabin logs, corrals, mine props, and railroad ties. The logs are too small and knotty to yield milled lumber.

Pinyon (piñon) pine

Pinus edulis Engelm.

Key Features: Needles are two per fascicle and 1 to 2 inches long. Cones are short with relatively few scales that bear bean-sized nuts.

Range and Ecology: These pines are almost entirely restricted to lower slopes, reaching the lower timberline of the Four Corners states. In Huerfano and Las Animas counties, they are the pinyon of the P-J formation. The trees have a low, broad, and irregular crown. By casual observation, they appear similar to the junipers with which they grow—especially when observed at 60 miles per hour. At the upper end of their elevational range, the P-J formation becomes nearly pure pinyon.

Trunks and branches are curved and slow growing, reaching a few hundred years old and producing pitch-laden wood. The oldest specimen in Colorado, found west of Meeker, has been growth-ring dated at about 900 years.

Pinyon pine has been one of the crucially important food plants for humans of this region since at least the Colorado archaic period (2000 to 9000 years before present) as attested by archeological findings. It continues to be collected among people native to this pine's range, who value this nutritional food source. The nuts, when eaten roasted or raw, provide as many calories and amino acids as steak. This species has numerous other uses, including medicinal and adhesive use of the pitch and as a component of dye recipes. The dense, pitch-laden wood is choice fuel, burning slowly and hot. At Mesa Verde and neighboring sites, the species is very common on mesa tops. Cones and nuts mature in about 18 months and can be harvested in August. However, high production, being partially dependent on climatic conditions, occurs only every half-dozen years.

Limber pine

Pinus flexilis E. James

Key Features: Needles are five to a fascicle, up to 3 inches long, and are forward pointing, producing densely tufted branches. Needles always lack the white waxy spots found on bristlecone pine needles.

Range and Ecology: The species ranges from southern Canada to California, to the eastern ranges of Colorado, New Mexico, and Arizona. They are documented from Huerfano County. The elevational range shows wide variation. As with bristlecone pine, they occur on exposed high elevation sites near timberline and therefore are never common. I have seen the species growing naturally as low as 8600 feet in Tracy Canyon (FR410). Never do they form large, gregarious stands. Whenever a small pine is seen growing out of a crack in a boulder, it is almost always this species.

These trees grow low and stocky, 10 to 18 feet tall, and up to 2 feet in diameter when mature. Such a specimen may be 200 to 300 years old. The bark is thick and has small surface scales. In the wind, the densely needled branches give off a whistling sound. As the name implies, the twigs are very flexible and capable of being tied in knots. Their irregular, gnarled mature form makes them highly photographed wherever found. Economically, they are too remote, crooked, and scarce to be valuable.

Mountain ponderosa pine, western yellow pine

Pinus ponderosa Douglas ex Lawson & C. Lawson var. *scopulorum* Engelm.

Key Features: Needles two to three to a fascicle and usually longer than 4 inches (10 cm), conspicuously longer than any other area pine (Fig. 79).

Range and Ecology: Widespread from southern British Columbia, south to California, and east to the eastern Rockies. Outlying populations also occur in the western Dakotas, Nebraska, New Mexico, Arizona, and into central Mexico. Our specimens, in both counties, are found at mid elevations. In mature stands, the large trunks are well spaced, giving an inviting savannah- or park-like aspect with a grassy or shrubby understory.

Ponderosas may grow in almost pure stands or are associated with pinyon pine at lower elevations, or with white or Douglas fir at higher elevations. The trees are not fussy about substrate but do need well-drained soil. They tolerate variable acidity (limestone to basaltic soils), clay to sand, or poorly formed soils. The best stands are on good soils of gently rolling terrain.

This species forms among the largest of Rocky Mountain forest trees, some specimens achieving a height of 125 feet and 30 inches in diameter. Their lifespan may be longer than 300 years here. In the Northwest, they grow much

Fig. 79 (above): Branch ending of ponderosa pine. Its needles are the longest among local pines.

Fig. 80 (left): Mature ponderosa pine bark, at least 75 years old.

older and larger. Bark of younger trees is dark and furrowed. In specimens older than 70 to 100 years, the bark is transformed into broad, orangish plates that are quite distinctive (Fig. 80). If you smell the mature bark, you can detect a distinctive pleasant odor variously said to resemble butterscotch or vanilla. Structurally, the bark is adapted to fire resistance. It has pitch-laden layers of loose cells that explode away from the tree when steam forms or when ignited by fast-moving ground fire.

Among pines, this species is the foremost timber tree of the West, growing well over millions of square miles. Historically, Flagstaff, Arizona, took its name from a ponderosa pine pole erected there on July 4, 1876. It became a landmark on the trail between Santa Fe and San Francisco. John Muir has said, "Of all the pines, this one gives forth the finest music of the winds."

DOUGLAS FIR, GENUS *PSEUDOTSUGA* (ONE SPECIES)

Rocky Mountain Douglas fir

Pseudotsuga menziesii (Mirb.) Franco var. *glauca* (Mayr) Franco

Key Features: Needles are flat and fir-like, up to 1.25 inches long. Unlike all other North American conifers, cones of this genus have elongated bracts emerging from between scales that are conspicuously three pointed.

Range and Ecology: Widespread from central British Columbia continuously down the Pacific Coast to central California. Eastern populations are scattered through the central and eastern Rockies to central Mexico. The species follows the central Sangre de Cristos and Spanish Peaks, occurring higher than ponderosa pine. It mixes in the lower subalpine zone with Engelmann spruce. Our specimens grow best in more open forest, are less tolerant of shade, and are more tolerant of drought. Although not documented here, the species is present in both counties.

The West Coast stands consist of giant trees with compact crowns, mast-like trunks, and dark blue-green needles. In 1825, the Scottish botanical explorer David Douglas found one that was 227 feet tall and 48 feet in circumference. Such a tree has almost no taper and is branch free for more than 100 feet. In the Northwest, where rainfall reaches 100 feet annually, specimens may live a thousand years; that is, if they can hide from chain saws. The Rocky Mountain subspecies is much smaller than the coastal version but can still be called majestic where it persists in a favorable location. On old trees, the bark is deeply furrowed and thick.

Commercially, this species has been hugely valuable as a timber source. Lumber barons at the turn of the 20th century ripped them from the coastal forests with all possible dispatch. Their great, strong logs made large wood structures possible, such as the huge roof beams of the Mormon Tabernacle in Salt Lake City. To this day, the wood is favored for laminated arches for field houses and other large spans. Thousands of miles of railroad track were laid on Douglas fir ties, and many telephone messages are supported en route by Douglas fir utility poles. Its historically important contributions form a very long list, and it remains the best conifer wood for anything from construction to interior finish carpentry.

CHAPTER 11

WOODY FLOWERING PLANTS (TREES, SHRUBS, AND WOODY CREEPERS)

Note: This checklist includes all species mapped in Flora of North America Editorial Committee (1993), Kartesz and Meacham (1999), and other sources, as occurring in or near this area. In some instances, included species are not documented. The point of including *possible* species is to alert those interested to look out for new occurrences and range extensions. Plants on the Colorado noxious weed list are marked [☐].

Field identification notes are included, but note that some species are not easily identified without access to technical keys.

Species found in Huerfano County are marked [H]; those found in Las Animas County are marked [L]; and those found in both counties are marked [HL].

MAPLES, FAMILY ACERACEAE (ONE GENUS)

Box elder

Acer negundo L. var. *negundo*

Key Features: Small or large tree. Only maple with compound leaves, three, five, or seven leaflets, occasionally resembling poison ivy but arranged oppositely (i.e., two leaves inserted on opposite sides of the same point on the stem). Twigs are green.

Range and Ecology: Widespread and weedy in North America. [L]

Dwarf maple, Rocky Mountain maple

Acer glabrum Torr.

Key Features: Small tree. Leaves opposite, palmately lobed or three parted.

Range and Ecology: Wide ranging in the West, usually in moist valleys in mountains. [HL]

SUMAC, FAMILY ANACARDIACEAE (ONE GENUS)

Skunkbrush, three-leafed sumac, squaw brush

Rhus trilobata Nutt. var. *pilosissima* Engelm. and *R. trilobata* var. *trilobata*

Key Features: Shrub. Twigs short-hairy. Leaves opposite, trifoliate, shallow-lobed or coarse-toothed. Petiolules short or absent. Leaves have a strong musky odor.

Range and Ecology: Mid elevations, in patches at woody borders. [HL]

BARBERRIES, FAMILY BERBERIDACEAE (ONE GENUS)

Red Oregon grape

Mahonia haematocarpa (Wooton) Fedde

Key Features: Creeping woody plant. Leaves alternate, compound with five to nine leaflets with terminal leaflet with five to seven spiny teeth. Flowers large, yellow. Berries purple.

Range and Ecology: Mostly southern (New Mexico, Arizona). Mid-elevation woods. Las Animas County collection isolated at north end of range.

Oregon grape

Mahonia repens (Lindl.) G. Don

Key Features: Creeping woody plant. Leaves alternate, five to seven leaflets with terminal leaflet. Leaves coarse-toothed with fine points. Flowers small, yellow. Berries purple.

Range and Ecology: Mid-elevation woods. [H]

ALDERS, BIRCHES, FAMILY BETULACEAE (ONE GENUS)

Water birch, brown birch

Betula occidentalis Hook.

Key Features: Small or large tree. Leaves alternate, ovate, toothed. Trunk reddish brown, smooth, with horizontal streaks. Bark peels horizontally. Fruit in catkins.

Range and Ecology: Common in western mountains and plains near water. Low to mid elevations. [HL]

Thinleaf alder, mountain alder

Alnus incana (L.) Moench subsp. *tenuifolia* (Nutt.) Breitung

Key Features: Shrub. Leaves irregularly or double-toothed. Bark dark. Fruit broad, short "cone" (catkin).

Range and Ecology: Low to mid elevations. Common in western plains and mountains, near water. Farther west, but it may be here.

ELDERBERRY, FAMILY CAPRIFOLIACEAE (FOUR GENERA)

Four-line honeysuckle

Lonicera involucrata (Richardson) Banks ex Spreng. var. *involucrata*

Key Features: Shrub. Leaves simple, opposite, ovate to elliptic. Margins entire. Flowers in pairs, elongate, yellow. Berries purple with red bracts subtending.

Range and Ecology: Widespread in the West. Open areas and margins. [HL]

Black elder

Sambucus nigra L. subsp. *canadensis* (L.) Bolli, and *S. nigra* subsp. *cerulea* (Raf.) Bolli

Key Features: Shrub. Leaves opposite, compound, five to 11 broad leaflets, with terminal leaflet, fine-toothed. Flowers in terminal clusters, white. Fruits dark blue, berry-like.

Range and Ecology: Widespread in North America. Mid elevations, open areas, and margins in Las Animas County.

Elderberry, red elder

Sambucus racemosa L. subsp. *racemosa*

Key Features: Leaves opposite, compound with terminal leaflet. Leaves toothed. Fruits blue, white waxy.

Range and Ecology: Widespread in western North America, protected woody areas, often near streams. [HL]

Common snowberry

Symphoricarpos albus (L.) S. F. Blake

Key Features: Shrub. Leaves opposite, simple, broad-elliptic with entire margins. Flowers white.

Range and Ecology: Northeastern United States and Mountain West. Open areas in Las Animas County.

Mountain snowberry

Symphoricarpos oreophilus A. Gray

Key Features: Shrub. Leaves opposite, simple, broad-elliptic, entire margins. Flowers white-pink, in axillary pairs. Fruit white.

Range and Ecology: Western, widespread. Range ends eastward at Great Plains. Canyons and valleys of Huerfano and Las Animas counties.

Western snowberry

Symphoricarpos occidentalis Hook.

Key Features: Shrub. Leaves opposite, simple, broad-elliptic. Margins entire. Flower stalks in leaf axils, several clusters per stalk.

Range and Ecology: Upper central United States. Huerfano and Las Animas counties are near range limit. Open areas.

Wayfaring tree

Viburnum lantana L.

Key Features: Shrub. Leaves opposite, simple, large-toothed. Flowers in terminal panicles.

Range and Ecology: Very scattered range. Upper Midwest and four western counties from Colorado to Montana. Las Animas County as scattered specimens in open areas.

BOXLEAF, FAMILY CELASTRACEAE (ONE GENUS)

Oregon boxleaf

Paxistima myrsinites (Pursh) Raf.

Key Features: Shrub, densely branched. Leaves ovate, shiny green, pointed-toothed, length:width ratio 2:1. Flowers maroon.

Range and Ecology: Mid elevation in mountains. [L]

DOGWOODS, FAMILY CORNACEAE (ONE GENUS)

Red osier dogwood

Cornus sericea L. subsp. *sericea*

Key Features: Twigs red, leaves opposite, broad-elliptic, margins smooth. Buds with two scales. Fruits white, fleshy.

Range and Ecology: Widespread in North America. Documented just west of here; seen by author in Huerfano County. In mountains at mid elevations, along moist trails.

RUSSIAN OLIVE, BUFFALO BERRY, FAMILY ELAEAGNACEAE (TWO GENERA)

Buffalo berry, soap berry

Shepherdia canadensis (L.) Nutt.

Key Features: Shrub. Leaves opposite, ovate, entire. Leaves covered with brown or silvery hairs. Flowers when leaf buds open. Berries red.

Range and Ecology: Widespread in West, scattered in Midwest and East. [HL]

Russian olive

Elaeagnus angustifolia L.

Key Features: Leaves alternate, long-elliptic, silvery green. Twigs silver-white.

Range and Ecology: Asian, introduced as windbreak plant. Invasive weed (□ in New Mexico). Not reported, but likely here.

HEATHS, FAMILY ERICACEAE (TWO GENERA)

Bear berry, kinnikinnick

Arctostaphylos uva-ursi (L.) Spreng.

Key Features: Small creeping woody plant. Leaves small, less than ½ inch, elliptic, entire. Flowers white, urn-shaped. Berries red.

Range and Ecology: Widespread in West, Upper Midwest, and Northeast. Huerfano County, in ground cover patches beneath montane forest canopies.

Grouseberry, whortleberry

Vaccinium myrtillus L.

Key Features: Creeping woody plant. Leaves broad-elliptic with tapering bases. Berries red.

Range and Ecology: Scattered in the West. In Huerfano County as understory ground cover.

LOCUST, FAMILY FABACEAE (THREE GENERA)

Fragrant mimosa

Mimosa borealis A. Gray

Key Features: Large shrub or small tree. Twigs slightly zigzag with flattened thorns. Leaves bipinnately compound, to 1.25 inches long. Three to five pairs of tiny leaflets per pinna. Inflorescences pink. Flowers ball-like. Fruit pods often constricted between seeds.

Range and Ecology: Texas, New Mexico to Colorado. On rocky hillsides. Las Animas County near northern range limit.

Honey mesquite

Prosopis glandulosa Torr. var. *glandulosa*

Key Features: Small or large shrub or tree. Twigs very thorny. Leaves forked, pinnately compound. Leaflets ½ to 1 inch long. Flowers in racemes, cream-colored. Fruits long brown constricted pods.

Range and Ecology: Southwestern; the subspecies restricted to New Mexico and Colorado and scattered eastward to Texas and Oklahoma. Rare in Las Animas County, the northern range limit.

New Mexico locust

Robinia neomexicana Gray var. *neomexicana*

Key Features: Leaves alternate, compound, nine to 21 hair-tipped leaflets. No terminal leaflet. Thorns in pairs at leaf scars. Flowers in racemes, pea-like, pink.

Range and Ecology: Mountains and plains of Colorado, New Mexico, west Texas, southeast Nevada. Las Animas and Huerfano counties at eastern range limit.

OAKS, FAMILY FAGACEAE (ONE GENUS)

Gambel oak, Rocky Mountain white oak

Quercus gambelii Nutt.

Key Features: Shrub-like or small tree. Trunk well-developed, of hard wood. Buds clustered at branch ends. Leaves deeply lobed with lobes rounded. Acorns present.

Range and Ecology: Four Corners states. Huerfano and Las Animas counties at eastern limit. Forms large hillside thickets at low to mid elevations.

BUCKTHORNS, FAMILY RHAMNACEAE (ONE GENUS)

Buckbrush

Ceanothus fendleri A. Gray

Key Features: Low shrub. Branches bluish gray. Leaves small, elliptic, trinerved, with small teeth. Flowers in showy white clusters. Fruit a capsule.

Range and Ecology: Widespread in North America. Mid to high elevations in Las Animas County.

SERVICEBERRY, PLUM, CHOKECHERRY, MOUNTAIN MAHOGANY, FAMILY ROSACEAE (SEVEN GENERA)

Serviceberry, shadbush

Amelanchier alnifolia (Nutt.) Nutt. ex M. Roem. var. *alnifolia*

Key Features: Leaves opposite, simple, broad oval, teeth distal or hardly present. Flowers with five separated white petals. Berries blue.

Range and Ecology: Rockies, southeastern range limit. [L]

Mountain mahogany

Cercocarpus montanus Raf.

Key Features: Shrub. Leaves small, distal half rounded, distal-toothed; basal half wedge-pointed toward base. Fruits small with elongated, curled, feather-like stalks.

Range and Ecology: Four Corners states and California, Texas near eastern limit. Common in shrub zone of mountain slopes. [HL]

Rock spiraea, ocean spray

Holodiscus dumosus (Nutt. ex Hook.) A. Heller

Key Features: Shrub. Leaves small, coarsely toothed. Basal teeth largest at halfway point. Flowers in large clusters.

Range and Ecology: Four Corners states and Nevada, Wyoming, and Texas. [L]

Mountain ninebark

Physocarpus monogynus (Torr.) J. M. Coult.

Key Features: Shrub. Leaves broad, palmate-lobed, and toothed. Flowers large, white, with five separated petals.

Range and Ecology: Rockies front, plus Arizona, Nevada, and New Mexico. [HL]

Atlantic ninebark

Physocarpus opulifolius (L.) Maxim. var. *intermedius* (Rydb.) B. L. Rob.

Key Features: Shrub. Leaves fine-toothed and with shallow lateral lobes. Flowers small and white.

Range and Ecology: Great Plains states and East. Las Animas County is western limit.

American plum

Prunus americana Marshall

Key Features: Small tree. Bark smooth. Leaves alternate, long (1 to 4 inches), pointed, margin often doubly serrate. Fruits fleshy, less than 1-inch diameter, red or yellow with one seed.

Range and Ecology: Widespread in North America. Common in Mountain West on plains or mountain slopes. [L]

Fire cherry

Prunus pensylvanica L. f.

Key Features: Small tree, trunk copper-red, smooth, with horizontal streaks. Twigs red and smooth. Leaves alternate, ovate-elliptic, fine-toothed. Flowers white, with five petals, in clusters. Fruits red berries.

Range and Ecology: Mostly northeastern, Las Animas County near southwestern range limit.

Common chokecherry, western chokecherry

Prunus virginiana L. var. *melanocarpa* (A. Nelson) Sarg.

Key Features: Small tree. Leaves broad, 1 to 4 inches long, finely and singly toothed. Small fruits, dark blue-purple to black.

Range and Ecology: Widespread in North America in wooded areas. [HL]

Prickly rose

Rosa acicularis Lindl. subsp. *sayi* (Schwein.) W. H. Lewis

Key Features: Shrub. Leaves compound with terminal leaflet. Leaflets three to seven, small, fine-toothed. Flowers white.

Range and Ecology: Rockies and Upper Midwest. [L]

Prairie rose

Rosa arkansana Porter

Key Features: Shrub. Stems thorny. Leaves compound in seven to nine toothed leaflets. Leaflets with teeth or entire. Distal leaf largest. Flowers pink.

Range and Ecology: Midwest to Rockies front. Las Animas County is near southwestern range limit.

Wood's rose

Rosa woodsii Lindl. var. *ultramontana* (S. Watson) Jeps.

Key Features: Shrub. Stems thorny. Leaves compound with seven to nine leaflets. Flowers rose or fuchsia. Fruits red-orange hips.

Range and Ecology: Western. Stops eastward at Great Plains. [HL]

Raspberry, thimbleberry

Rubus deliciosus Torr.

Key Features: Shrub. Leaves palmate, trilobed, irregularly toothed, 1 to 2 inches long. Flowers solitary, white, five petals, yellow stamens. Fruits aggregate and dark purple.

Range and Ecology: Limited range, Wyoming, Colorado, and northeastern New Mexico. Las Animas County is near southern range limit.

Red raspberry

Rubus idaeus L. subsp. *strigosus* (Michx.) Focke

Key Features: Shrub. Stem with numerous fine, hard hairs. Young twigs red-violet. Leaves compound, leaflets three to five and terminal leaflet. Distal and basal leaflets largest. Fruits red, compound, berry-like.

Range and Ecology: Western, widespread, to Upper Midwest and Northeast. Rare in plains. [HL]

HOPTREE, RUES, FAMILY RUTACEAE (ONE GENUS)

Hoptree

Ptelea trifoliata L. subsp. *pallida* (Greene) V. L. Bailey

Key Features: Shrub or small tree. Leaves trifoliate, long petioles, leaflets sessile at petiole summit. Finely or scarcely toothed. Flowers small, white. Fruits winged and dry.

Range and Ecology: Southwestern, scattered: Arizona, Utah, New Mexico. In Colorado: Las Animas County.

POPLARS AND WILLOWS, FAMILY SALICACEAE (TWO GENERA)

Cottonwoods and aspens

Populus L.

The largest broad-leaved trees of the arid West, they are usually found in bottomlands near water. Trunks may reach more than 3 feet in diameter and have deeply furrowed bark.

Cottonwood (of hybrid origin)

Populus ×acuminata Rydb.

Key Features: Tall tree. Old bark brown, deeply furrowed. Leaves ovate or elliptic with tapering leaf bases. Toothed but few at basal margin. Length: width ratio 2:1 to 3:1.

Range and Ecology: Scattered in mountain states. [HL]

Narrow-leaf cottonwood or poplar

Populus angustifolia E. James

Key Features: Tree. Leaves narrowest of any cottonwood, resembling those of willows. Length:width ratio 4:1 to 5:1.

Range and Ecology: Throughout the central mountain region. Moist areas in Huerfano and Las Animas counties.

Eastern cottonwood, plains cottonwood

Populus deltoides W. Bartram ex Marshall subsp. *monilifera* (Aiton) Eckenw. (includes *P. sargentii* Dode)

Key Features: Tall tree. Trunk often of great girth. Old bark deeply furrowed. Leaves deltoid, coarsely serrate, thick and shiny.

Range and Ecology: Great Plains to Rockies front and Upper Midwest. [HL]

Quaking aspen, trembling aspen

Populus tremuloides Michx.

Key Features: Small to large tree at mid elevations. Bark smooth and light gray-green, persisting for decades. Older trees develop gray-brown furrowed bark. Leaves deltoid with pointed tip, thin and not usually shiny, 1 to 3 inches long, with small marginal rounded teeth. Petioles laterally flattened, causing leaves to oscillate or shimmer in breezes.

Range and Ecology: Common throughout the Rockies, Upper Midwest, and Northeast. Throughout the tree zone of lower to upper slopes (Fig. 81). Growing in nearly pure groves by root suckering, the trees may get large and old. An aspen clone growing in Utah's Wasatch Mountains may be the world's heaviest living organism. It is composed of 47,000 stems arising from an underground root system estimated to weigh 6500 tons.

Willows (*Salix* L.)

The genus is easily sight-recognized, but technical keys and flowering material are often required for species identification. Locally, most are small trees or shrubs with narrow, often elliptic leaves. Shrubs along banks of streams or lakes are usually willows.

Peachleaf willow

Salix amygdaloides Andersson

Key Features: Large or small tree. Twigs gray to yellow. Leaves up to 13 cm long, thin, lance-ovate, and widest for all the local species, finely toothed, yellow-green, paler beneath. Length:width ratio 4:1 to 5:1.

Range and Ecology: West to northeastern United States. Along streams at intermediate elevations in Huerfano and Las Animas counties.

Bebb's willow, beaked willow

Salix bebbiana Sarg.

Key Features: Shrub or small tree. Twigs reddish, yellow-brown to dark brown, growth irregular with dieback stubs. Leaves elliptic, up to 6 cm long, tips and base rounded, entire to finely toothed, shiny or dull green, paler beneath, length:width ratio 3:1 to 4:1.

Range and Ecology: West, Upper Midwest, and Northeast. Found along

Fig. 81: Aspens and some conifers growing on slopes that were once burned within the past century. Aspens are frequently a pioneer second-growth species. Under some conditions, they can persist for centuries because of their long-lived root system that replaces senescent stems.

streams or in moist meadows. Most frequent in the spruce-fir zone but also at lower elevations. [HL]

Sage willow

Salix candida Fluggé ex Willd.

Key Features: Twigs yellow to brown. Leaves narrow-elliptic, dark or light green, whitish beneath.

Range and Ecology: Rare, mountain states, Upper Midwest, and Northeast. Huerfano County is in southern end of scattered Rocky Mountain range.

Drummond's willow

Salix drummondiana Barratt ex Hook.

Key Features: Twigs purplish. Leaves dark green and thickened, whitish beneath, margins appearing entire. Length:width ratio 5:1.

Range and Ecology: Mountain states. Huerfano County is in the eastern margin of intermountain western range.

Missouri willow, diamond willow

Salix eriocephala Michx.

Key Features: Twigs yellow. Leaves narrow, shiny, fine-toothed. Length: width ratio 5:1 to 7:1. Pairs of large rounded stipules at leaf base.

Range and Ecology: Eastern, scattered in Colorado, rare in Huerfano County.

Coyote willow

Salix exigua Nutt.

Key Features: Twigs reddish. Leaves narrow, lance-shaped, fine-toothed, dark green to yellow-green, paler beneath. Length:width ratio 6:1 to 10:1.

Range and Ecology: Western, extending east a little beyond the western Great Plains. Often in lowlands and stabilizes sandbars. Huerfano and Las Animas counties are near eastern end of range.

Sandbar willow

Salix interior Rowlee (perhaps a subspecies of *S. exigua*)

Key Features: Twigs reddish. Leaves narrow. Length:width ratio 6:1 to 7:1.

Range and Ecology: Eastern and Midwestern, scattered in Rocky Mountain counties including Huerfano and Las Animas.

Yellow willow

Salix lutea Nutt.

Key Features: Similar to *Salix eriocephala*. Shrub to 15 feet tall. Twigs yellow to brown. Leaves elliptic, fine-toothed. Length:width ratio 3:1 to 4:1.

Range and Ecology: Scattered west of Mississippi River to far West. Stream banks to floodplains. Isolated occurrence in Huerfano County.

Park willow, mountain willow

Salix monticola Bebb

Key Features: Thicket-forming shrub. Twigs green. Leaves elliptic, finely toothed, dark shiny green, lighter green beneath. Inflorescence red. Length: width ratio 3:1 to 4:1.

Range and Ecology: Southern Rockies, Four Corners states, and southern Wyoming. Near water courses, mid elevations to subalpine in Huerfano County.

Snow willow

Salix nivalis Hook.

Key Features: Sub-shrub or prostrate. Twigs green. Leaves nearly round, length:width ratio 1:1 to 1.5:1, entire or finely toothed, shiny dark green above.

Range and Ecology: Scattered in intermountain west above timberline. [H]

Planeleaf willow

Salix planifolia Pursh

Key Features: Small tree. Twigs shiny, red-brown to black. Leaves fine-toothed, 2 to 3 inches long, green, paler beneath. Length:width ratio 3:1 to 4:1.

Range and Ecology: Upper Midwest and scattered western states. Stream banks and meadows. [L]

SOAPBERRIES, FAMILY SAPINDACEAE (One Genus)

Soapberry

Sapindus saponaria L. var. *drummondii* (Hook. & Arn.) L. D. Benson

Key Features: Large or small tree. Old bark rough with exfoliating plates. Leaves alternate, compound with ca. 16 leaflets and no terminal leaflet. Leaflets elliptic. Length:width ratio 3:1 to 4:1. Berries yellow.

Range and Ecology: Southern United States. Rare in Las Animas County, the northern range limit.

ELMS, FAMILY ULMACEAE (Two Genera)

Hackberry

Celtis occidentalis L.

Key Features: Tall tree. Trunk and branches with warty knobs. Leaves toothed, to 6 inches long with uneven (asymmetric) leaf base. Leaves toothed in distal ⅔ of blade. Berries red.

Range and Ecology: Widespread in North America, at low elevations here (< 6000 feet). [L]

Sugarberry

Celtis laevigata Willd. var. *reticulata* (Torr.) L. D. Benson

Key Features: Large or small tree. Bark warty. Leaves simple, ovate, to 3 to 4 inches long, bases uneven or even. Margins finely or scarcely toothed. Berries red-yellow.

Range and Ecology: Scattered western. Las Animas County, lower elevations.

Siberian elm

Ulmus pumila L.

Key Features: Tree. Leaves alternate, to 3 inches long, margins finely single-toothed, leaf base uneven.

Range and Ecology: Introduced and originally planted. Often escaped and naturalized in most states. Not reported for Huerfano and Las Animas counties but probably here at lower elevations.

CHAPTER 12

NON-WOODY FLOWERING PLANTS

THIS LIST has been compiled from various regional floras. In addition, sources included volumes of the *Flora of North America* that have been so far completed, plus the CD-ROM database *Synthesis of the North American Flora* by Kartesz and Meacham (1999). The most comprehensive source for botanical names was the *Checklist of Vascular Plants of the Southern Rocky Mountain Region* by Snow (2009). There are competing naming concepts, and I relied on no single source; in places, I inserted my own additions. Families are listed alphabetically, as are genera and species within families. Using floras with keys and illustrations, one can use the following lists to narrow down the species possibilities. The inclusion of subspecies (subsp.) and varieties (var.) will be of value to some users.

A NOTE ON NAMES

The non-italicized, often abbreviated person-names following the plant genus and species track the history of the species in the literature. More importantly, they are tied to specific descriptions and specimens that allow botanists to organize this constantly evolving field. I have attempted to supply at least one "common" name per genus in the hope that it will be helpful to some users. However, unlike latinized names, their boundaries are often vague. Common names may be the same for different species, or several may be used for one species. Some species are too rare to have a common name. If you care to really identify a plant, the often colorful common names are not sufficient. I hope the ones chosen for this list will be helpful.

ACCURACY

The majority of plants on these lists have been documented from the region. For the remainder, their ranges include the two counties and they should be expected to occur here. Botanists who write floras depend on collections that are properly prepared and deposited in recognized museums. If fieldwork is insufficient, many species will not be documented as part of a county flora. For this interesting region, it turned out to be premature to expect an accurate, fully documented flora.

I discovered that tree species, which would seem hard to miss, are about 30% under-reported. The *quality* of work done to date is not the problem. There are simply too few experts with detailed knowledge of the families of organisms to provide total coverage of all counties. The flora of the United States, comprising approximately 22,000 species, is not well documented for the majority of the more than 3000 counties.

For this list, I have erred on the side of inclusion. It is a snapshot of opinion in an ongoing enterprise. My hope is that local interested naturalists will be stimulated to take up the study of the flora and fauna of this interesting region. Even a modest amount of fieldwork can contribute substantially to our knowledge by adding new records and clarifying identifications.

COLLECTING PLANTS

Many of the names in this list conjure up images of edible, medicinal, and flavoring plants. Collecting plants for such uses requires expert knowledge, and I recommend against collecting plants from our flora. A plant that closely resembles a food plant may be poisonous. If a plant is on the rare and endangered species list, collecting it is illegal.

A FEW STATISTICS

The following table lists basic counts for species, at the subspecies and variety level, for the non-woody plants of the region.

Occurring in Huerfano County	577
Occurring in Las Animas County	837
Occurring in both counties	390
In Huerfano County only	187
In Las Animas County only	447
Total species listed	1024

Considering the larger size of Las Animas County, the distribution is about as expected. The Huerfano County flora is about 20% unique for the two-county area, and the Las Animas County flora is 48% unique. This guide is meant to cover the region in the vicinity of the Spanish Peaks, but the list doubtless covers numerous species in these counties on the plains to the east. From the sparse documentation at hand, we have no basis to exclude species occurring just outside this guide's defined territory.

The top 11 families of herbaceous flowering plants and their species numbers for the two counties are: Asteraceae, 206; Poaceae, 147; Fabaceae, 74; Brassicaceae, 41; Cyperaceae, 35; Ranunculaceae, 27; Amaranthaceae, 27; Plantaginaceae, 25; Polygonaceae, 25; Onagraceae, 23; and Apiaceae, 23. Considering the tentative nature of this list, no further analysis will be attempted.

Plants considered rare or extirpated by some authorities are so marked. Finally, the U.S. Bureau of Land Management, the USDA Natural Resources Conservation Service, and many state agencies list noxious and invasive weeds. These are defined as aggressively colonizing, non-native plants that threaten the integrity of natural ecosystems. Unfortunately, they are increasing nationwide. They are also marked in this list.

Species that are found in Huerfano County are marked [H]; those found in Las Animas County are marked [L]; those found in both counties are marked [HL]; and those appearing on the noxious weed list of Colorado or adjacent states are marked [☐].

FAMILY ACERACEAE (SEE CHAPTER 11, WOODY FLOWERING PLANTS)

FAMILY ADOXACEAE

Muskroot

Adoxa moschatellina L. [H]

FAMILY ALISMATACEAE

Water plantain

Alisma triviale Pursh [L]

Arrowhead

Sagittaria cuneata E. Sheld. [L]
Sagittaria graminea Michx. var. *graminea* [L, rare]
Sagittaria latifolia Willd. [L]

ALLIACEAE, ONIONS

Onion

Allium cernuum Roth [L]
Allium geyeri S. Watson var. *geyeri* [HL]
Allium textile A. Nelson & J. F. Macbr. [HL]

FAMILY AMARANTHACEAE

Pigweed
Amaranthus arenicola I. M. Johnst. [L]
Amaranthus blitoides S. Watson [HL]
Amaranthus palmeri S. Watson [L]
Amaranthus retroflexus L. [HL]

Saltbush
Atriplex canescens (Pursh) Nutt. var. *canescens* [HL]
Atriplex hortensis L. [H] □
Atriplex powellii S. Watson [H]
Atriplex rosea L. [L] □

Goosefoot, pigweed
Chenopodium berlandieri Moq. [L]
Chenopodium capitatum (L.) Asch. [H]
Chenopodium cycloides A. Nelson [L, rare]
Chenopodium fremontii S. Watson var. *fremontii* [HL]
Chenopodium graveolens Willd. [L]
Chenopodium incanum (S. Watson) A. Heller var. *incanum* [L]
Chenopodium leptophyllum (Moq.) Nutt. ex S. Watson [L]
Chenopodium overi Aellen [L]
Chenopodium pratericola Rydb. [L]
Chenopodium rubrum L. [H]
Chenopodium simplex (Torr.) Raf. [L]
Chenopodium watsonii A. Nelson [HL]

Winged pigweed
Cycloloma atriplicifolium (Spreng.) J. M. Coult. [L]

Snakecotton
Froelichia gracilis (Hook.) Moq. [L]

Mexican fireweed
Kochia scoparia (L.) Schrad. [L] □

Winterfat
Krascheninnikovia lanata (Pursh) A. Meeuse & A. Smit [HL]

Povertyweed
Monolepis nuttalliana (Schult.) Greene [H]

Russian thistle
Salsola tragus L. [HL] □

Woolly honeysweet
Tidestromia lanuginosa (Nutt.) Standl. [H]

FAMILY ANACARDIACEAE (SEE CHAPTER 11, WOODY FLOWERING PLANTS)

FAMILY APIACEAE

Angelica
Angelica grayi (J. M. Coult. & Rose) J. M. Coult. & Rose {L}

Cutleaf water parsnip
Berula erecta (Huds.) Coville {HL}

Spotted water parsnip or hemlock
Cicuta maculata L. {H}

Hemlock parsley
Conioselinum scopulorum (A. Gray) J. M. Coult. & Rose {L}

Poison hemlock
Conium maculatum L. {L} □

Mountain springparsley
Cymopterus acaulis (Pursh) Raf. {HL}
Cymopterus montanus Torr. & A. Gray {HL}

Indian parsley
Cymopterus anisatus A. Gray {L}

Carrot
Daucus carota L. {L} □

Whiskbroom parsley
Harbouria trachypleura (A. Gray) J. M. Coult. & Rose {L}

Cow parsnip
Heracleum sphondylium L. var. *lanatum* (Michx.) Dorn {HL}

Wild lovage
Ligusticum porteri J. M. Coult. & Rose var. *porteri* {H}

Biscuitroot, desert parsley
Lomatium foeniculaceum (Nutt.) J. M. Coult. & Rose var. *foeniculaceum* {L}

Leafy wildparsley
Musineon divaricatum (Pursh) Nutt. {L}

False Indian parsley
Neoparrya lithophila Mathias {H, rare}

Alpine parsley
Oreoxis alpina (A. Gray) J. M. Coult. & Rose {L}

Sweet cicely
Osmorhiza chilensis Hook. & Arn. [L]
Osmorhiza depauperata Phil. [HL]
Osmorhiza longistylis (Torr.) DC. [L]

Cowbane
Oxypolis fendleri (A. Gray) A. Heller [HL]

Parsnip
Pastinaca sativa L. [H] ☐

Woodroot
Podistera eastwoodiae (J. M. Coult. & Rose) Mathias & Constance [H]

False mountain parsley
Pseudocymopterus montanus (A. Gray) J. M. Coult. & Rose [HL]

FAMILY APOCYNACEAE

Dogbane
Apocynum androsaemifolium L. [HL]
Apocynum cannabinum L. [L]

Milkweed
Asclepias asperula (Decne.) Woodson var. *asperula* [HL]
Asclepias engelmanniana Woodson [L]
Asclepias involucrata Engelm. ex Torr. [L, rare]
Asclepias latifolia (Torr.) Raf. [HL]
Asclepias macrotis Torr. [L, rare]
Asclepias oenotheroides Schltdl. & Cham. [L, rare]
Asclepias pumila (A. Gray) Vail [L]
Asclepias speciosa Torr. [HL]
Asclepias subverticillata (A. Gray) Vail [HL]
Asclepias tuberosa L. subsp. *interior* Woodson [L]
Asclepias uncialis Greene [HL, rare]
Asclepias viridiflora Raf. [L]

Climbing milkweed
Funastrum crispum (Benth.) Schltr. [L, rare]

FAMILY ARACEAE

Duckweed
Lemna minor L. [L]

FAMILY ASPLENIACEAE (SEE CHAPTER 9, FERNS)

FAMILY ASTERACEAE, ASTERS

Yarrow
Achillea millefolium L. var. *lanulosa* (Nutt.) Piper [HL]

Fragrant snakeroot
Ageratina herbacea (A. Gray) R. M. King & H. Rob. [L]

Goat chicory
Agoseris aurantiaca (Hook.) Greene var. *aurantiaca* [H]
Agoseris glauca (Pursh) Raf. [H]

Ragweed
Ambrosia acanthicarpa Hook. [L]
Ambrosia confertiflora DC. [L]
Ambrosia psilostachya DC. [L]
Ambrosia tomentosa Nutt. [L]
Ambrosia trifida L. var. *trifida* [L]

Pearly everlasting
Anaphalis margaritacea (L.) Benth. & Hook. [L]

Pussytoes
Antennaria anaphalioides Rydb. [H]
Antennaria media Greene [H]
Antennaria microphylla Rydb. [H]
Antennaria neglecta Greene [H]
Antennaria parvifolia Nutt. [HL]
Antennaria rosea Greene [HL]

Burdock
Arctium minus Bernh. [L] □

Heartleaf arnica
Arnica cordifolia Hook. [HL] (Fig. 82)
Arnica mollis Hook. [H]

Sagebrush, wormwood
Artemisia bigelovii A. Gray [L]
Artemisia campestris L. var. *pacifica* (Nutt.) M. Peck [L]
Artemisia carruthii Alph. Wood ex Carruth. [HL]
Artemisia dracunculus L. [L]
Artemisia filifolia Torr. [L]
Artemisia franserioides Greene [L]
Artemisia frigida Willd. [HL]
Artemisia ludoviciana Nutt. var. *incompta* (Nutt.) Cronquist [L]
Artemisia ludoviciana Nutt. var. *ludoviciana* [HL]
Artemisia parryi A. Gray [L]

Groundsel tree
Baccharis wrightii A. Gray [L]

False goldfields, yellow ragweed
Bahia dissecta (A. Gray) Britton [HL]

Fig. 82: Heart-leaved arnica (*Arnica cordifolia*) carpets the forest along shaded woodland trails on montane slopes.

Greeneyes
Berlandiera lyrata Benth. [L]

Beggarticks
Bidens bigelovii A. Gray [L]
Bidens bipinnata L. [L]
Bidens cernua L. [L] ☐
Bidens tenuisecta A. Gray [H]

False boneset
Brickellia brachyphylla (A. Gray) A. Gray [L]
Brickellia californica (Torr. & A. Gray) A. Gray var. *californica* [L]
Brickellia eupatorioides (L.) Shinners var. *chlorolepis* (Wooton & Standl.) B. L. Turner [L]
Brickellia grandiflora (Hook.) Nutt. var. *grandiflora* [L]

Dustymaiden
Chaenactis douglasii (Hook.) Hook. & Arn. [H]

Rose heath
Chaetopappa ericoides (Torr.) G. L. Nesom [HL]

Rabbitbrush
Chrysothamnus depressus Nutt. [L]
Chrysothamnus greenei (A. Gray) Greene [H]
Chrysothamnus vaseyi (A. Gray) Greene [H]

Chrysothamnus viscidiflorus (Hook.) Nutt. var. *lanceolatus* (Nutt.) Greene [H]
Chrysothamnus viscidiflorus (Hook.) Nutt. var. *stenophyllus* (A. Gray) H. M. Hall [H]

Thistle
Cirsium arvense (L.) Scop. [H] ☐
Cirsium ochrocentrum A. Gray [L]
Cirsium pallidum Wooton & Standl. [L]
Cirsium parryi (A. Gray) Petr. subsp. *parryi* [L]
Cirsium scopulorum (Greene) Cockerell ex Daniels [L]
Cirsium undulatum (Nutt.) Spreng. [L]

Horseweed
Conyza canadensis (L.) Cronquist [L]

Hawksbeard
Crepis nana Richardson [H, rare]
Crepis runcinata (E. James) Torr. & A. Gray [HL]

Fetid marigold
Dyssodia papposa (Vent.) Hitchc. [HL]

Purple coneflower
Echinacea angustifolia DC. var. *angustifolia* [L]

Engelmann's daisy
Engelmannia pinnatifida A. Gray ex Nutt. [L]

Rubber rabbitbrush
Ericameria nauseosa (Pall. ex Pursh) G. L. Nesom & G. I. Baird [HL]
Ericameria parryi (A. Gray) G. L. Nesom & G. I. Baird var. *affinis* (A. Nelson) G. L. Nesom & G. I. Baird [H]
Ericameria parryi (A. Gray) G. L. Nesom & G. I. Baird var. *parryi* [HL]

Fleabane
Erigeron bellidiastrum Nutt. [L]
Erigeron canus A. Gray [HL]
Erigeron colomexicanus A. Nelson [L]
Erigeron coulteri Porter [HL]
Erigeron divergens Torr. & A. Gray [HL]
Erigeron elatior (A. Gray) Greene [L]
Erigeron engelmannii A. Nelson var. *engelmannii* [H]
Erigeron eximius Greene [HL]
Erigeron flagellaris A. Gray [HL]
Erigeron formosissimus Greene [HL]
Erigeron glabellus Nutt. [L]
Erigeron lonchophyllus Hook. [H]
Erigeron peregrinus (Banks ex Pursh) Greene subsp. *callianthemus* (Greene) Cronquist [L]
Erigeron pumilus Nutt. subsp. *pumilus* [HL]
Erigeron simplex Greene [H]
Erigeron speciosus (Lindl.) DC. [L]

Erigeron strigosus Muhl. ex Willd. var. *strigosus* [L]
Erigeron subtrinervis Rydb. ex Porter & Britton var. *subtrinervis* [HL]
Erigeron vetensis Rydb. [L]

Blanket flower
Gaillardia aristata Pursh [HL]
Gaillardia pinnatifida Torr. [HL]

Gumweed
Grindelia acutifolia Steyerm. [HL]
Grindelia decumbens Greene [L]
Grindelia inornata Greene [L]
Grindelia nuda Alph. Wood var. *aphanactis* (Rydb.) G. L. Nesom [HL]
Grindelia nuda Alph. Wood var. *nuda* [HL]
Grindelia revoluta Steyerm. [H]
Grindelia squarrosa (Pursh) Dunal var. *squarrosa* [HL]

Snakeweed
Gutierrezia sarothrae (Pursh) Britton & Rusby [HL]

Sneezeweed
Helenium microcephalum DC. [L, rare]

Dwarf sunflower
Helianthella parryi A. Gray [L]
Helianthella quinquenervis (Hook.) A. Gray [HL]

Sunflower
Helianthus annuus L. [HL]
Helianthus nuttallii Torr. & A. Gray [L]
Helianthus petiolaris Nutt. var. *petiolaris* [HL]
Helianthus rigidus (Cass.) Desf. [H]
Helianthus tuberosus L. [L]

Oxeye
Heliopsis helianthoides (L.) Sweet var. *occidentalis* (T. R. Fisher) Steyerm. [HL]

Spiny aster
Herrickia horrida Wooton & Standl. [L, rare]

Wingpetal
Heterosperma pinnatum Cav. [L, rare]

Golden aster
Heterotheca canescens (DC.) Shinners [L]
Heterotheca pumila (Greene) Semple [H]
Heterotheca villosa (Pursh) Shinners var. *foliosa* (Nutt.) V. L. Harms [HL]
Heterotheca villosa (Pursh) Shinners var. *minor* (Hook.) Semple [H]
Heterotheca villosa (Pursh) Shinners var. *nana* (A. Gray) Semple [HL]
Heterotheca villosa (Pursh) Shinners var. *villosa* [L]

Low alpine hawkweed
Hieracium gracile Hook. var. *gracile* [H]

Old plainsman
Hymenopappus filifolius Hook. var. *cinereus* (Rydb.) I. M. Johnst. [HL]
Hymenopappus filifolius Hook. var. *parvulus* (Greene) B. L. Turner [L]
Hymenopappus flavescens A. Gray var. *flavescens* [L]
Hymenopappus tenuifolius Pursh [L]

Bitterweed, orange sneezeweed
Hymenoxys hoopesii (A. Gray) Bierner [HL]
Hymenoxys richardsonii (Hook.) Cockerell [HL]

Marsh elder
Iva axillaris Pursh [H]

Careless weed
Iva xanthiifolia Nutt. [HL]

Lettuce
Lactuca oblongifolia Nutt. [HL]
Lactuca serriola L. [HL] □

Blazing star
Liatris ligulistylis (A. Nelson) K. Schum. [HL]
Liatris punctata Hook. [HL]

Skeleton plant
Lygodesmia juncea (Pursh) D. Don ex Hook. [HL]

Tansy aster
Machaeranthera bigelovii (A. Gray) Greene var. *bigelovii* [HL]
Machaeranthera canescens (Pursh) A. Gray var. *aristata* (Eastw.) B. L. Turner [H]
Machaeranthera pinnatifida (Hook.) Shinners var. *glaberrima* (Rydb.) B. L. Turner & R. L. Hartm. [HL]
Machaeranthera tanacetifolia (Kunth) Nees [HL]

Tarweed
Madia glomerata Hook. [L] □

Blackfoot
Melampodium leucanthum Torr. & A. Gray [HL]

Prairie dandelion
Nothocalais cuspidata (Pursh) Greene [L]

Scotch thistle
Onopordum acanthium L. [L] □
Onopordum tauricum Willd. [H] □

False goldenrod
Oonopsis engelmannii (A. Gray) Greene [L]

Oonopsis foliosa (A. Gray) Greene [L]
Oonopsis foliosa (A. Gray) Greene var. *monocephala* (A. Nelson) Kartesz & Gandhi [L, rare]

Ragwort, Groundsel
Packera crocata (Rydb.) W. A. Weber & Á. Löve [L]
Packera fendleri (A. Gray) W. A. Weber & Á. Löve [HL]
Packera neomexicana (A. Gray) W. A. Weber & Á. Löve var. *mutabilis* (Greene) W. A. Weber & Á. Löve [L]
Packera plattensis (Nutt.) W. A. Weber & Á. Löve [L]
Packera pseudaurea (Rydb.) W. A. Weber & Á. Löve var. *flavula* (Greene) Trock & T. M. Barkley [H]
Packera streptanthifolia (Greene) W. A. Weber & Á. Löve [H]
Packera tridenticulata (Rydb.) W. A. Weber & Á. Löve [HL]

Spanish needles
Palafoxia rosea (Bush) Cory [L]
Palafoxia sphacelata (Nutt. ex Torr.) Cory [L]

American or alpine feverfew
Parthenium alpinum (Nutt.) Torr. & A. Gray [L, rare]

Cinchweed
Pectis angustifolia Torr. var. *angustifolia* [L]

Mountain tail-leaf
Pericome caudata A. Gray [HL]

False bahia
Picradeniopsis oppositifolia (Nutt.) Rydb. ex Britton [HL]

Rabbit-tobacco
Pseudognaphalium canescens (DC.) Anderb. [L]

Paperflower
Psilostrophe bakeri Greene [H]
Psilostrophe tagetina (Nutt.) Greene var. *tagetina* [L]

Mexican hat
Ratibida columnifera (Nutt.) Wooton & Standl. [L]
Ratibida tagetes (E. James) Barnhart [HL]

Goldenglow
Rudbeckia hirta L. var. *pulcherrima* Farw. [HL]
Rudbeckia laciniata L. subsp. *ampla* (A. Nelson) W. A. Weber [HL]

Manyflower false threadleaf
Schkuhria multiflora Hook. & Arn. [L]

Groundsel, ragwort
Senecio amplectens A. Gray var. *amplectens* [HL]
Senecio atratus Greene [H]

Senecio bigelovii A. Gray var. *hallii* A. Gray [HL]
Senecio eremophilus Richardson var. *kingii* Greenm. [H]
Senecio flaccidus Less. var. *flaccidus* [L]
Senecio integerrimus Nutt. var. *exaltatus* (Nutt.) Cronquist [HL]
Senecio spartioides Torr. & A. Gray var. *fremontii* (Torr. & A. Gray) Greenm. ex L. O. Williams [L]
Senecio spartioides Torr. & A. Gray var. *multicapitatus* (Greenm. ex Rydb.) S. L. Walsh [H]
Senecio taraxacoides (A. Gray) Greene [HL]
Senecio triangularis Hook. [L]
Senecio wootonii Greene [HL]

Goldenrod
Solidago canadensis L. var. *gilvocanescens* Rydb. [H]
Solidago canadensis L. var. *scabra* (Muhl. ex Willd.) Torr. & A. Gray [HL]
Solidago gigantea Aiton [HL]
Solidago missouriensis Nutt. [L]
Solidago multiradiata Aiton var. *scopulorum* A. Gray [H]
Solidago parryi (A. Gray) Greene [HL]
Solidago rigida L. var. *rigida* [L]
Solidago simplex Kunth subsp. *simplex* [HL]
Solidago velutina DC. [L]
Solidago wrightii A. Gray var. *adenophora* S. F. Blake [L]

Sowthistle
Sonchus uliginosus M. Bieb. [H] □

Prairie skeleton plant or desert straw
Stephanomeria pauciflora (Torr.) A. Nelson var. *pauciflora* [HL]

Wild aster
Symphyotrichum campestre (Nutt.) G. L. Nesom [L]
Symphyotrichum falcatum (Lindl.) G. L. Nesom var. *falcatum* [L]
Symphyotrichum fendleri (A. Gray) G. L. Nesom [L]
Symphyotrichum laeve (L.) Á. Löve & D. Löve var. *geyeri* (A. Gray) G. L. Nesom [L]
Symphyotrichum novae-angliae (L.) G. L. Nesom [L] (Fig. 83)
Symphyotrichum oblongifolius (Nutt.) G. L. Nesom [L]
Symphyotrichum porteri (A. Gray) G. L. Nesom [L]

Dandelion
Taraxacum officinale F. H. Wigg. [HL]

Horsebrush
Tetradymia canescens DC. [H]

Bitterweed
Tetraneuris acaulis (Pursh) Greene var. *acaulis* [HL]
Tetraneuris acaulis (Pursh) Greene var. *caespitosa* A. Nelson [H]

Fig. 83: Purple asters (*Aster* species) along the North Fork Trail.

Tetraneuris brandegeei (Porter ex A. Gray) K. F. Parker [H]
Tetraneuris scaposa (DC.) Greene var. *scaposa* [L]

Navajo tea, Hopi tea
Thelesperma filifolium (Hook.) A. Gray var. *intermedium* (Rydb.) Shinners [HL]
Thelesperma megapotamicum (Spreng.) Kuntze [L]
Thelesperma subnudum A. Gray var. *subnudum* [HL]

Prickly leaf
Thymophylla aurea (A. Gray) Greene ex Britton & A. Br. var. *aurea* [L]

Townsend's aster, or daisy
Townsendia eximia A. Gray [HL]
Townsendia exscapa (Richardson) Porter [HL]
Townsendia grandiflora Nutt. [L]

Salsify, goat's-beard
Tragopogon dubius Scop. [HL] □ (Fig. 84)
Tragopogon porrifolius L. [L] □

Crownbeard
Verbesina encelioides (Cav.) Benth. & Hook. f. ex A. Gray subsp. *exauriculata* (B. L. Rob. & Greenm.) J. R. Coleman [HL]

Fig. 84: Goatsbeard daisy (*Tragopogon dubius*). An attractive non-native meadow plant.

False goldeneye
Viguiera multiflora (Nutt.) S. F. Blake var. *nevadensis* (A. Nelson) S. F. Blake [HL]

Cocklebur
Xanthium spinosum L. [L] ☐
Xanthium strumarium L. var. *canadense* (Mill.) Torr. & A. Gray [HL] ☐

Rocky Mountain zinnia
Zinnia grandiflora Nutt. [HL]

FAMILY BERBERIDACEAE

Creeping mahonia, Oregon grape
Mahonia haematocarpa (Wooton) Fedde [L]
Mahonia repens (Lindl.) G. Don [H]

FAMILY BETULACEAE (SEE CHAPTER 11, WOODY FLOWERING PLANTS)

FAMILY BORAGINACEAE

Cat's eye
Cryptantha cinerea (Greene) Cronquist var. *jamesii* (Torr.) Cronquist [L]
Cryptantha crassisepala (Torr. & A. Gray) Greene [L]

Cryptantha fendleri (A. Gray) Greene [HL]
Cryptantha minima Rydb. [HL]
Cryptantha thyrsiflora (Greene) Payson [HL]

Houndstongue
Cynoglossum officinale L. [HL] □

Beggar's lice
Hackelia floribunda (Lehm.) I. M. Johnst. [HL]

Stickseed
Lappula occidentalis (S. Watson) Greene var. *cupulata* (Gray) Higgins [HL]
Lappula occidentalis (S. Watson) Greene var. *occidentalis* [HL]

Puccoon
Lithospermum incisum Lehm. [HL]
Lithospermum multiflorum Torr. ex A. Gray [HL]

Bluebells
Mertensia ciliata (James ex Torr.) G. Don var. *ciliata* [H]
Mertensia franciscana A. Heller [HL]
Mertensis lanceolata (Pursh) A. DC. [HL]

Marbleseed
Onosmodium molle Michx. var. *occidentale* (Mack.) I. M. Johnst. [L]

FAMILY BRASSICACEAE

Alyssum, madwort
Alyssum alyssoides (L.) L. [H]
Alyssum simplex Rudolphi [H]

Rock cress
Arabis hirsuta (L.) Scop. var. *pycnocarpa* (M. Hopkins) Rollins [L]

Yellow rocket
Barbarea vulgaris W. T. Aiton [L] □

Rock cress
Boechera divaricarpa (A. Nelson) Á. Löve & D. Löve [H]
Boechera drummondii (A. Gray) Á. Löve & D. Löve [HL]
Boechera fendleri (S. Watson) W. A. Weber var. *fendleri* [L]
Boechera fendleri (S. Watson) W. A. Weber var. *spatifolia* (Rydb.) Dorn [HL]

False flax
Camelina microcarpa Andrz. ex DC. [HL] □

Shepherd's purse
Capsella bursa-pastoris (L.) Medik. [HL] □

Bittercress
Cardamine cordifolia A. Gray [H]

Hare's ear mustard

Conringia orientalis (L.) Dumort. [L]

Tansy mustard

Descurainia incana (Bernh. ex Fisch. & C. A. Mey.) Dorn subsp. *incisa* (Engelm. ex A. Gray) Kartesz & Gandhi [H]
Descurainia incana (Bernh. ex Fisch. & C. A. Mey.) Dorn var. *incana* [HL]
Descurainia pinnata (Walter) Britton [HL]
Descurainia sophia (L.) Webb ex Prantl [L] □

Whitlow wort

Draba aurea Vahl ex Hornem. [H]
Draba crassa Rydb. [H, rare]
Draba grayana (Rydb.) C. L. Hitchc. [H, rare]
Draba helleriana Greene var. *helleriana* [H]
Draba rectifructa C. L. Hitchc. [H, rare]
Draba reptans (Lam.) Fernald [L]
Draba smithii Gilg & O. E. Schulz [L, rare]
Draba streptocarpa A. Gray [HL]

Wallflower

Erysimum asperum (Nutt.) DC. [HL]
Erysimum capitatum (Douglas ex Hook.) Greene var. *argillosum* (Greene) R. J. Davis [HL] (Fig. 85)

Fig. 85: Wallflower (*Erysimum capitatum*). Petals vary from light yellow to deep orange on the same plant.

Erysimum cheiranthoides L. [H]
Erysimum inconspicuum (S. Watson) MacMill. [L]

Pepperwort
Lepidium densiflorum Schrad. [HL]

Bladderpod
Lesquerella calcicola Rollins [HL]
Lesquerella montana (A. Gray) S. Watson [HL]
Lesquerella ovalifolia Rydb. ex Britton var. *ovalifolia* [HL]

Cress
Nasturtium officinale R. Br. [L]

Wild candytuft
Noccaea montana (L.) F. K. Mey. var. *montana* [HL]

Cress
Rorippa sinuata (Nutt.) Hitchc. [L]

Charlock
Sinapis arvensis L. [L] □

Tumblemustard, hedgemustard
Sisymbrium altissimum L. [HL]

Prince's-plume
Stanleya pinnata (Pursh) Britton [HL]

Plainsmustard
Thelypodiopsis linearifolius (A. Gray) Al-Shehbaz [L]

Thelypody
Thelypodium wrightii A. Gray subsp. *oklahomensis* Al-Shehbaz [L]

Pennycress
Thlaspi arvense L. [HL]

FAMILY CACTACEAE, CACTI

Spinystar, foxtail
Coryphantha vivipara (Nutt.) Britton & Rose var. *arizonica* (Engelm. ex W. H. Brewer & S. Watson) W. T. Marshall [HL]

Hedgehog
Echinocereus coccineus Engelm. var. *coccineus* [HL]
Echinocereus reichenbachii (Terscheck) Britton & Rose subsp. *perbellus* (Britton & Rose) N. P. Taylor [L, rare]
Echinocereus viridiflorus Engelm. var. *viridiflorus* [HL]

Prickly pear, cholla
Opuntia imbricata (Haw.) DC. var. *imbricata* [L]
Opuntia macrorhiza Engelm. var. *macrorhiza* [HL]

Opuntia phaeacantha Engelm. [L]
Opuntia polyacantha Haw. var. *polyacantha* [HL]

Pincushion
Pediocactus simpsonii (Engelm.) Britton & Rose var. *simpsonii* [H]

FAMILY CAMPANULACEAE

Harebell, bellflower
Campanula parryi A. Gray var. *parryi* [HL]
Campanula rapunculoides L. [H, □ weedy]
Campanula rotundifolia L. [HL]

Cardinal flower
Lobelia cardinalis L. [L]

FAMILY CANNABACEAE

Hops
Humulus lupulus L. [L]

FAMILY CAPRIFOLIACEAE (SEE ALSO CHAPTER 11, WOODY FLOWERING PLANTS)

Twinflower
Linnaea borealis L. [H]

FAMILY CARYOPHYLLACEAE

Sandwort
Arenaria fendleri A. Gray var. *fendleri* [HL]
Arenaria hookeri Nutt. [HL]
Arenaria hookeri Nutt. subsp. *pinetorum* (A. Nelson) W. A. Weber [L]
Arenaria hookeri Nutt. var. *hookeri* [L]
Arenaria lanuginosa (Michx.) Rohrb. subsp. *saxosa* (A. Gray) Maguire [H]

Chickweed
Cerastium arvense L. [HL]
Cerastium beeringianum Cham. & Schltdl. var. *capillare* Fernald & Wiegand [H]
Cerastium brachypodum (Engelm. ex A. Gray) B. L. Rob. [H]
Cerastium fontanum Baumg. subsp. *vulgare* (Hartm.) Greuter & Burdet [L]

Alpine sandwort, stichwort
Minuartia obtusiloba (Rydb.) House [H]

Wood sandwort, grove sandwort
Moehringia lateriflora (L.) Fenzl [HL]

Nailwort
Paronychia jamesii Torr. & A. Gray [L]
Paronychia sessiliflora Nutt. [HL]

Tuber starwort
Pseudostellaria jamesiana (Torr.) W. A. Weber & R. L. Hartm. [L]

Catchfly
Silene acaulis (L.) Jacq. var. *subacaulescens* (F. N. Williams) Fernald & H. St. John [H]
Silene dioica (L.) Clairv. [H]
Silene drummondii Hook. var. *drummondii* [L]
Silene scouleri Hook. subsp. *hallii* (S. Watson) C. L. Hitchc. & Maguire [L]

Chickweed, starwort
Stellaria longifolia Muhl. ex Willd. [HL]
Stellaria longipes Goldie var. *longipes* [H]

FAMILY CERATOPHYLLACEAE

Coontail, hornwort
Ceratophyllum demersum L. [H]

FAMILY CLEOMACEAE

Spider flower, beeplant
Cleome serrulata Pursh [HL]

Clammyweed
Polanisia dodecandra (L.) DC. [HL]
Polanisia jamesii (Torr. & A. Gray) Iltis [L]

FAMILY CLUSIACEAE

St. John's wort
Hypericum formosum Kunth var. *nortoniae* (M. E. Jones) C. L. Hitchc. [L]

FAMILY COMMELINACEAE

Dayflower
Commelina dianthifolia Delile [L]
Commelina erecta L. [L]

Spiderwort
Tradescantia occidentalis (Britton) Smyth var. *occidentalis* [L]

FAMILY CONVOLVULACEAE

Hedge bindweed, false bindweed
Calystegia sepium (L.) R. Br. subsp. *angulata* Brummitt [HL]

Bindweed
Convolvulus arvensis L. [HL] ☐
Convolvulus equitans Benth. [L] ☐

Dodder
Cuscuta cuspidata Engelm. [L]
Cuscuta indecora Choisy var. *neuropetala* (Engelm.) Hitchc. [L]

Dwarf morning-glory
Evolvulus nuttallianus J. A. Schultes [L]

Morning-glory
Ipomoea leptophylla Torr. [L]

FAMILY CORNACEAE (SEE CHAPTER 11, WOODY FLOWERING PLANTS)

FAMILY CRASSULACEAE

Rosewort, queen's crown
Sedum integrifolium (Raf.) A. Nelson [L]
Sedum rhodanthum A. Gray [HL]

Stonecrop
Sedum lanceolatum Torr. [L]

FAMILY CROSSOSOMATACEAE

Plains greasebush
Forsellesia planitierum Ensign [L, rare]

FAMILY CUCURBITACEAE

Buffalo gourd
Cucurbita foetidissima Kunth [HL]

Bur cucumber
Cyclanthera dissecta (Torr. & A. Gray) Arn. [L]

Wild cucumber
Echinocystis lobata (Michx.) Torr. & A. Gray [H]

FAMILY CUPRESSACEAE (SEE CHAPTER 10, GYMNOSPERMS)

FAMILY CYPERACEAE

Sedge
Carex aquatilis Wahlenb. [HL]
Carex arapahoensis Clokey [HL]
Carex aurea Nutt. [L]
Carex bella L. H. Bailey [HL]
Carex brevior (Dewey) Mack. [L]
Carex douglasii Boott [HL]
Carex duriuscula C. A. Mey. [L]
Carex ebenea Rydb. [HL]

Carex emoryi Dewey [L]
Carex geophila Mack. [L]
Carex gravida L. H. Bailey var. *lunelliana* F. J. Herm. [L, rare]
Carex heteroneura W. Boott var. *chalciolepis* (T. Holm) F. J. Herm. [H]
Carex hystericina Muhl. ex Willd. [L]
Carex interior L. H. Bailey [L]
Carex microptera Mack. [HL]
Carex nova L. H. Bailey [HL]
Carex occidentalis L. H. Bailey [L]
Carex pellita Muhl. ex Willd. [HL]
Carex pensylvanica Lam. var. *digyna* Boeck. [L]
Carex praegracilis W. Boott [HL]
Carex scopulorum T. Holm [L]
Carex siccata Dewey [L]
Carex subfusca W. Boott [H]
Carex utriculata W. Boott [HL]
Carex vallicola Dewey var. *vallicola* [L]
Carex vulpinoidea Michx. var. *vulpinoidea* [L]

Umbrella sedge
Cyperus lupulinus (Spreng.) Marcks subsp. *lupulinus* [L]
Cyperus schweinitzii Torr. [L]

Spikerush
Eleocharis palustris (L.) Roem. & Schult. var. *palustris* [HL]

Cottongrass
Eriophorum angustifolium Honck. [L]
Eriophorum gracile W. D. J. Koch var. *gracile* [L, rare]

Bulrush
Schoenoplectus americanus (Pers.) Volkart ex Schinz & R. K. Heller [L]
Schoenoplectus maritimus (L.) Lye [L]
Schoenoplectus tabernaemontani (C. C. Gmel.) Palla [L]
Scirpus pallidus (Britton) Fernald [L]

FAMILY DENNSTAEDTIACEAE (SEE CHAPTER 9, FERNS)

FAMILY DRYOPTERIDACEAE (SEE CHAPTER 9, FERNS)

FAMILY ELEAGNACEAE (SEE CHAPTER 11, WOODY FLOWERING PLANTS)

FAMILY EQUISETACEAE (SEE CHAPTER 9, FERNS)

FAMILY ERICACEAE (SEE ALSO CHAPTER 11, WOODY FLOWERING PLANTS)

Bearberry
Arctostaphylos uva-ursi (L.) Spreng. [H]

Pipsissewa, prince's pine
Chimaphila umbellata (L.) W. P. C. Barton subsp. *occidentalis* (Rydb.) Hultén [HL]

Wood nymph
Moneses uniflora (L.) A. Gray [H]

Pinesap
Monotropa hypopitys L. [L]

Sidebells
Orthilia secunda (L.) House [HL]

Pinedrops
Pterospora andromedia Nutt. [L]

Wintergreen
Pyrola chlorantha Sw. [H]
Pyrola minor L. [H]

Whortleberry
Vaccinium myrtillus L. [H]

FAMILY EUPHORBIACEAE

Tall silverbush
Argythamnia mercurialina (Nutt.) Müll. Arg. var. *mercurialina* [L]

Spurge, sandmat
Chamaesyce fendleri (Torr. & A. Gray) Small [HL]
Chamaesyce geyeri (Engelm.) Small var. *geyeri* [L]
Chamaesyce glyptosperma (Engelm.) Small [L]
Chamaesyce lata (Engelm.) Small [L]
Chamaesyce missurica (Raf.) Shinners [L]
Chamaesyce serpens (Kunth) Small [L]
Chamaesyce stictospora (Engelm.) Small [L]

Croton
Croton texensis (Klotzsch) Müll. Arg. var. *texensis* [L]

Spurge
Euphorbia brachycera Engelm. [HL]
Euphorbia dentata Michx. [L]
Euphorbia marginata Pursh [HL]

Noseburn
Tragia ramosa Torr. [HL]

FAMILY FABACEAE (SEE ALSO CHAPTER 11, WOODY FLOWERING PLANTS)

Lead plant, indigo bush

Amorpha canescens Pursh [L]
Amorpha fruticosa L. [HL]
Amorpha nana Nutt. [L, rare]

Milkvetch

Astragalus agrestis Douglas ex G. Don [H]
Astragalus alpinus L. var. *alpinus* [HL]
Astragalus bisulcatus (Hook.) A. Gray [HL]
Astragalus bodinii E. Sheld. [L, rare]
Astragalus canadensis L. var. *canadensis* [HL]
Astragalus ceramicus E. Sheld. var. *filifolius* (A. Gray) F. J. Herm. [L]
Astragalus crassicarpus Nutt. var. *paysonii* (E. H. Kelso) Barneby [HL]
Astragalus drummondii Douglas ex Hook. [HL]
Astragalus flexuosus Douglas ex G. Don [HL]
Astragalus gracilis Nutt. [L]
Astragalus hallii A. Gray var. *hallii* [L]
Astragalus laxmannii Jacq. var. *robustior* (Hook.) Barneby & S. L. Welsh [HL]
Astragalus lonchocarpus Torr. [HL]
Astragalus lotiflorus Hook. [L]
Astragalus miser Douglas var. *oblongifolius* (Rydb.) Cronquist [H]
Astragalus missouriensis Nutt. [HL]
Astragalus mollissimus Torr. var. *mollissimus* [L]
Astragalus nuttallianus DC. [L]
Astragalus parryi A. Gray [H]
Astragalus pectinatus (Douglas ex Hook.) Douglas ex G. Don [L]
Astragalus puniceus Osterh. var. *puniceus* [L]
Astragalus racemosus Pursh var. *longisetus* M. E. Jones [L]
Astragalus scopulorum Porter [L]
Astragalus shortianus Nutt. [HL]
Astragalus tenellus Pursh [L]

Prairie clover

Dalea aurea Nutt. ex Pursh [L]
Dalea candida Michx. ex Willd. var. *oligophylla* (Torr.) Shinners [HL]
Dalea enneandra Nutt. [L]
Dalea formosa Torr. [L]
Dalea jamesii (Torr.) Torr. & A. Gray [HL]
Dalea purpurea Vent. var. *purpurea* [HL]
Dalea tenuifolia (A. Gray) Shinners [L]
Dalea villosa (Nutt.) Spreng. var. *villosa* [L]

Tick trefoil, bundleflower

Desmanthus cooleyi (Eaton) Trel. [HL]
Desmanthus illinoensis (Michx.) MacMill. ex B. L. Rob. & Fernald [L]

Licorice
Glycyrrhiza lepidota Pursh [HL]

Plains sweet-broom
Hedysarum boreale Nutt. subsp. *boreale* [L]

Rush pea
Hoffmannseggia drepanocarpa A. Gray [L]
Hoffmannseggia glauca (Ortega) Eifert [L] □

Wild pea, sweetpea
Lathyrus eucosmus Butters & H. St. John [HL]
Lathyrus laetivirens Greene ex Rydb. [HL]

Everlasting peavine
Lathyrus latifolius L. [L]
Lathyrus polymorphus Nutt. subsp. *incanus* (J. G. Sm. & Rydb.) Dorn [L]

Lupine, bluebonnet
Lupinus argenteus Pursh var. *argenteus* [HL] (Fig. 86)
Lupinus kingii S. Watson [H]

Fig. 86: Lupine (*Lupinus* sp.), with pale blue flowers growing near a rock wall at ca. 9000-feet elevation.

Lupinus plattensis S. Watson [HL]
Lupinus pusillus Pursh [HL]

Medick
Medicago lupulina L. [HL]
Medicago sativa L. [HL]

Sweet clover
Melilotus officinalis (L.) Pall. [HL]

Rocky Mountain loco, locoweed
Oxytropis lambertii Pursh var. *bigelovii* A. Gray [HL]
Oxytropis sericea Nutt. var. *sericea* [HL]

Prairie turnip
Pediomelum argophyllum (Pursh) J. W. Grimes [L]
Pediomelum hypogaeum (Nutt. ex Torr. & A. Gray) Rydb. var. *hypogaeum* [L]

Hog potato
Pomaria jamesii (Torr. & A. Gray) Walp. [L]

Scurfy pea, lemonweed
Psoralidium lanceolatum (Pursh) Rydb. [H]
Psoralidium tenuiflorum (Pursh) Rydb. [HL]

White loco, necklacepod
Sophora nuttalliana B. L. Turner [HL]

Golden banner
Thermopsis montana Nutt. var. *divaricarpa* (A. Nelson) Dorn [HL]
Thermopsis montana Nutt. var. *montana* [L]
Thermopsis rhombifolia (Nutt. ex Pursh) Richardson [HL]

Clover
Trifolium andinum Nutt. var. *andinum* [HL, rare]
Trifolium attenuatum Greene [HL]
Trifolium hybridum L. [H] □
Trifolium nanum Torr. [H]
Trifolium parryi A. Gray subsp. *parryi* [H]
Trifolium pratense L. [HL] □
Trifolium repens L. [HL] □
Trifolium wormskioldii Lehm. [H]

Vetch
Vicia americana Muhl. ex Willd. subsp. *minor* Hook. [HL]
Vicia ludoviciana Nutt. var. *ludoviciana* [L]

FAMILY FAGACEAE (SEE CHAPTER 11, WOODY FLOWERING PLANTS)

FAMILY FRANKENIACEAE

Seaheath
Frankenia jamesii Torr. ex A. Gray [L]

FAMILY GENTIANACEAE

Green gentian
Frasera coloradensis (C. M. Rogers) D. M. Post [L, rare]
Frasera speciosa Douglas ex Griseb. [L] (Fig. 87)

Gentian
Gentiana fremontii Torr. [L]
Gentiana parryi Engelm. [H]
Gentiana prostrata Haenke [H]

Dwarf gentian
Gentianella amarella (L.) Börner subsp. *acuta* (Michx.) J. M. Gillett [HL]

Fringed gentian
Gentianopsis barbellata (Engelm.) Iltis [H]

Fig. 87: Green gentian, or monument plant (*Frasera speciosa*). This plant blooms once, as infrequently as 20 to 60 years; sets seed; and dies.

Star gentian
Swertia perennis L. [L]

FAMILY GERANIACEAE

Storksbill
Erodium cicutarium (L.) L'Hér. ex Aiton subsp. *cicutarium* [HL] ☐

Cranesbill
Geranium atropurpureum A. Heller var. *atropurpureum* [H]
Geranium caespitosum E. James var. *fremontii* (Torr. ex A. Gray) Dorn [HL]
Geranium richardsonii Fisch. & Trautv. [HL]
Geranium viscosissimum Fisch. & C. A. Mey. var. *incisum* (Torr. & A. Gray) N. H. Holmgren [L]

FAMILY GROSSULARIACEAE

Gooseberry, Currant
Ribes aureum Pursh var. *aureum* [HL]
Ribes cereum Douglas [HL]
Ribes inerme Rydb. var. *inerme* [HL]
Ribes laxiflorum Pursh [L]
Ribes leptanthum A. Gray [L]
Ribes montigenum McClatchie [HL]
Ribes wolfii Rothr. [HL]

FAMILY HALORAGACEAE

Water milfoil
Myriophyllum sibiricum Kom. [H]

FAMILY HYDRANGEACEAE

Cliffbush
Jamesia americana Torr. & A. Gray var. *americana* [HL]

Mock orange
Philadelphus microphyllus A. Gray [L]

FAMILY HYDROCHARITACEAE

Waterweed
Elodea bifoliata H. St. John [L]
Elodea nuttallii (Planch.) H. St. John [L]

FAMILY HYDROPHYLLACEAE

Waterpod
Ellisia nyctelea (L.) L. [L]

Waterleaf
Hydrophyllum fendleri (A. Gray) A. Heller var. *fendleri* [HL]

Scorpionweed
Phacelia alba Rydb. [HL]
Phacelia denticulata Osterh. [HL]
Phacelia heterophylla Pursh subsp. *heterophylla* [H]

FAMILY HYPOXIDACEAE

Star grass
Hypoxis hirsuta (L.) Coville [L, rare]

FAMILY IRIDACEAE

Iris, blueflag
Iris missouriensis Nutt. [HL]

Blue-eyed grass
Sisyrinchium montanum Greene var. *montanum* [HL]

FAMILY JUNCACEAE

Rush
Juncus arcticus Willd. subsp. *littoralis* (Engelm.) Hultén [HL]
Juncus bufonius L. [HL]
Juncus drummondii E. Mey. [HL]
Juncus dudleyi Wiegand [L]
Juncus interior Wiegand [L]
Juncus longistylis Torr. [H]
Juncus mertensianus Bong. [L]
Juncus parryi Engelm. [H]
Juncus tenuis Willd. [L]
Juncus torreyi Coville [HL]

Woodrush
Luzula parviflora (Ehrh.) Desv. [H]

FAMILY JUNCAGINACEAE

Arrowgrass
Triglochin palustris L. [L]

FAMILY KRAMERIACEAE

Trailing ratany
Krameria lanceolata Torr. [L]

FAMILY LAMIACEAE

Giant hyssop
Agastache foeniculum Kuntze [HL, rare]

Wild basil
Clinopodium vulgare L. [L]

Dragonhead
Dracocephalum parviflorum Nutt. [HL]

Mock pennyroyal
Hedeoma drummondii Benth. [L]

Motherwort
Leonurus cardiaca L. [H]

Bugleweed, water horehound
Lycopus americanus Muhl. ex W. P. C. Barton [HL]

Horehound
Marrubium vulgare L. [HL]

Mint
Mentha arvensis L. [HL]
Mentha spicata L. [L]

Horsemint, beebalm
Monarda fistulosa L. var. *menthifolia* (Graham) Fernald [HL]
Monarda pectinata Nutt. [HL]

Selfheal
Prunella vulgaris L. subsp. *lanceolata* (W. P. C. Barton) Hultén [HL]

Sage
Salvia reflexa Hornem. [L]

Skullcap
Scutellaria brittonii Porter [HL]
Scutellaria galericulata L. [L]

Hedge nettle
Stachys palustris L. var. *pilosa* (Nutt.) Fernald [L]

Germander
Teucrium laciniatum Torr. [HL]

FAMILY LENTIBULARIACEAE

Common bladderwort
Utricularia vulgaris L. [L]

FAMILY LILIACEAE

Mariposa lily
Calochortus gunnisonii S. Watson var. *gunnisonii* [HL] (Fig. 88)

Sand lily
Leucocrinum montanum Nutt. ex A. Gray [L]

Wood lily
Lilium philadelphicum L. var. *andinum* Ker Gawl. [L, rare]

Alp lily
Lloydia serotina (L.) Rchb. subsp. *serotina* [H]

False Solomon's seal
Maianthemum racemosum (L.) Link var. *amplexicaule* (Nutt.) Dorn [HL]
Maianthemum stellatum (L.) Link [H]

Beargrass
Nolina texana S. Watson [L, rare]

Twistedstalk
Streptopus amplexifolius (L.) DC. var. *chalazatus* Fassett [H]

False hellebore, corn lily
Veratrum tenuipetalum A. Heller [H]

Fig. 88: Pink flowers of the mariposa lily (*Calochortus gunnisonii*), found above 9000 feet, and never common.

Yucca, soapweed, Spanish bayonet
Yucca baccata Torr. var. *baccata* [L]
Yucca baileyi Wooton & Standl. var. *baileyi* [H] (Fig. 89)
Yucca glauca Nutt. var. *glauca* [L]
Yucca neomexicana Wooton & Standl. [HL]

Death camus
Zigadenus elegans Pursh subsp. *elegans* [HL]
Zigadenus venenosus S. Watson var. *gramineus* (Rydb.) O. S. Walsh ex C. L. Hitchc. [HL]

Fig. 89: Yucca (*Yucca baileyi*). Infrequent on dry open meadows.

FAMILY LINACEAE

Flax
Linum australe A. Heller var. *australe* [L]
Linum lewisii Pursh var. *lewisii* [HL]
Linum puberulum (Engelm.) A. Heller [L]
Linum rigidum Pursh var. *rigidum* [L]

FAMILY LOASACEAE

Stickleaf, blazingstar
Mentzelia albicaulis (Douglas ex Hook.) Douglas ex Torr. & A. Gray [HL]
Mentzelia decapetala (Pursh ex Sims) Urb. & Gilg ex Gilg [L]
Mentzelia multiflora (Nutt.) A. Gray var. *multiflora* [HL]
Mentzelia nuda (Pursh) Torr. & A. Gray var. *nuda* [HL]
Mentzelia oligosperma Nutt. ex Sims [L]
Mentzelia reverchonii (Urb. & Gilg) H. J. Thomps. & Zavort. [L]
Mentzelia rusbyi Wooton [H]
Mentzelia speciosa Osterh. [L]

FAMILY MALVACEAE

Poppymallow
Callirhoe involucrata (Torr. & A. Gray) A. Gray [L]

Mallow
Malva neglecta Wallr. [HL]

Checkermallow
Sidalcea candida A. Gray var. *glabrata* C. L. Hitchc. [HL]
Sidalcea neomexicana A. Gray subsp. *neomexicana* [H]

Globemallow
Sphaeralcea angustifolia (Cav.) G. Don [HL]
Sphaeralcea coccinea (Nutt.) Rydb. [HL]

FAMILY MARSILIACEAE (SEE CHAPTER 9, FERNS)

FAMILY MARTYNIACEAE

Unicorn plant
Proboscidea louisianica Thell. subsp. *louisianica* [L]

FAMILY NAJADACEAE

Naiad
Najas guadalupensis (Spreng.) Magnus subsp. *guadalupensis* [L]

FAMILY NYCTAGINACEAE

Snowball, sand verbena
Abronia fragrans Nutt. ex Hook. [HL]

Trailing windmills
Allionia incarnata L. var. *incarnata* [L]

Four o'clock
Mirabilis glabra (S. Watson) Standl. [L]
Mirabilis hirsuta (Pursh) MacMill. [HL]
Mirabilis linearis (Pursh) Heimerl [HL]
Mirabilis multiflora (Torr.) A. Gray [HL]
Mirabilis nyctaginea (Michx.) MacMill. [L]
Mirabilis oxybaphoides (A. Gray) A. Gray [L]
Mirabilis rotundifolia (Greene) Standl. [L, rare]

Sand puffs
Tripterocalyx micranthus (Torr.) Hook. [HL]

FAMILY ONAGRACEAE

Sundrops
Calylophus hartwegii (Benth.) P. H. Raven subsp. *pubescens* (A. Gray) Towner & P. H. Raven [L]
Calylophus lavandulifolius (Torr. & A. Gray) P. H. Raven [HL]
Calylophus serrulatus (Nutt.) P. H. Raven [L]

Fireweed
Chamerion angustifolium (L.) Holub var. *canescens* (Alph. Wood) N. H. Holmgren & P. K. Holmgren [HL] (Fig. 90)

Willowherb
Epilobium brachycarpum C. Presl [L]
Epilobium ciliatum Raf. subsp. *ciliatum* [H]
Epilobium ciliatum Raf. subsp. *glandulosum* (Lehm.) Dorn [H]
Epilobium clavatum Trel. [H]
Epilobium hornemannii Rchb. subsp. *hornemannii* [HL]
Epilobium lactiflorum Hausskn. [H]

Beeblossom
Gaura coccinea Pursh [HL]
Gaura parviflora Douglas ex Lehm. [L]
Gaura villosa Torr. subsp. *villosa* [L]

Groundsmoke
Gayophytum diffusum Torr. & A. Gray var. *strictipes* (Hook.) Dorn [H]
Gayophytum ramosissimum Torr. & A. Gray [H]

Evening primrose
Oenothera albicaulis Pursh [HL]

Fig. 90: The purple-flowered fireweed (*Chamerion angustifolium*).

Oenothera caespitosa Nutt. [HL]
Oenothera coronopifolia Torr. & A. Gray [HL]
Oenothera elata Kunth subsp. *hirsutissima* (A. Gray ex S. Watson) Cronquist [HL]
Oenothera flava (A. Nelson) Garrett [H]
Oenothera harringtonii W. L. Wagner, Stockh. & W. M. Klein [HL, rare]
Oenothera latifolia (Rydb.) Munz [L]
Oenothera villosa Thunb. var. *strigosa* (Rydb.) Dorn [HL]

FAMILY OPHIOGLOSSACEAE (SEE CHAPTER 9, FERNS)

FAMILY ORCHIDACEAE

Calypso, fairy slipper
Calypso bulbosa (L.) Oakes var. *americana* (R. Br.) Luer [HL]

Spotted orchid
Coeloglossum viride (L.) Hartm. [L]

Coralroot
Corallorhiza striata Lindl. [HL]
Corallorhiza trifida Châtel. [HL]

Lady's slipper
Cypripedium parviflorum Salisb. var. *pubescens* O. W. Knight [L, rare]

Rattlesnake plantain
Goodyera oblongifolia Raf. [HL]
Goodyera repens (L.) R. Br. [L, rare]

Bog orchid
Platanthera aquilonis Sheviak [HL]
Platanthera dilatata (Pursh) Lindl. ex L. C. Beck var. *albiflora* (Cham.) Ledeb. [L]
Platanthera dilatata (Pursh) Lindl. ex L. C. Beck var. *dilatata* [L]
Platanthera purpurascens (Rydb.) Sheviak & W. F. Jenn. [HL]

Lady's tresses
Spiranthes romanzoffiana Cham. [H]

FAMILY OROBANCHACEAE

Indian paintbrush
Castilleja haydenii (A. Gray) Cockerell [H]
Castilleja integra A. Gray var. *integra* [HL]
Castilleja linariifolia Benth. [HL]
Castilleja lineata Greene [L, rare]
Castilleja miniata Douglas ex Hook. subsp. *miniata* [HL]
Castilleja occidentalis Torr. [H]
Castilleja sessiliflora Pursh [L]
Castilleja sulphurea Rydb. [H]

Blue-eyed Mary
Collinsia parviflora Lindl. [H]

Squawroot
Conopholis alpina Liebm. var. *mexicana* (A. Gray ex S. Watson) R. R. Haynes [L, rare]

Broomrape
Orobanche fasciculata Nutt. [HL]
Orobanche ludoviciana Nutt. var. *multiflora* (Nutt.) Beck [L]

Owl's clover
Orthocarpus luteus Nutt. [HL]

Owl's clover
Orthocarpus purpureoalbus A. Gray ex S. Watson [L]

Lousewort
Pedicularis canadensis L. subsp. *fluviatilis* (Heller) W. A. Weber [HL]
Pedicularis groenlandica Retz. [H]

Pedicularis parryi A. Gray subsp. *parryi* [H]
Pedicularis procera Adams ex Steven [HL]
Pedicularis racemosa Douglas ex Hook. subsp. *alba* Pennell [H]
Pedicularis sudetica Willd. subsp. *scopulorum* (A. Gray) Hultén [H]

FAMILY OXALIDACEAE

Wood sorrel

Oxalis dillenii Jacq. subsp. *dillenii* [L]
Oxalis violacea L. [HL]

FAMILY PAPAVERACEAE

Pricklypoppy

Argemone hispida A. Gray [L]
Argemone polyanthemos (Fedde) G. B. Ownbey [HL]
Argemone squarrosa Greene var. *squarrosa* [HL]

Fumewort, scrambled eggs

Corydalis aurea Willd. [HL]
Corydalis curvisiliqua Engelm. subsp. *occidentalis* (Engelm. ex A. Gray) W. A. Weber [HL]

Fumitory

Fumaria vaillantii Loisel. [L]

FAMILY PARNASSIACEAE

Grass of Parnassus

Parnassia palustris L. var. *montanensis* C. L. Hitchc. [L]

FAMILY PHRYMACEAE

Monkeyflower

Mimulus glabratus Kunth [HL]
Mimulus glabratus Kunth var. *jamesii* (Torr. & A. Gray ex Benth.) A. Gray [H]
Mimulus guttatus DC. [HL]
Mimulus rubellus Gray [H]

FAMILY PINACEAE (SEE CHAPTER 10, GYMNOSPERMS)

FAMILY PLANTAGINACEAE

Coraldrops

Besseya plantaginea (E. James) Rydb. [HL]

Toadflax

Linaria dalmatica (L.) Mill. subsp. *dalmatica* [L] ☐
Linaria vulgaris Mill. [HL] ☐

Beardtongue
Penstemon albidus Nutt. [L]
Penstemon angustifolius Nutt. ex Pursh var. *caudatus* (A. Heller) Rydb. [HL]
Penstemon auriberbis Pennell [HL]
Penstemon barbatus (Cav.) Roth var. *torreyi* (Benth.) A. Gray [HL]
Penstemon brandegii (Porter) Porter ex Rydb. [L]
Penstemon cobaea Nutt. [L]
Penstemon glaber Pursh var. *alpinus* (Torr.) A. Gray [L]
Penstemon gracilis Nutt. var. *gracilis* [HL]
Penstemon jamesii Benth. [L, rare]
Penstemon strictus Benth. [H]
Penstemon versicolor Pennell [L]
Penstemon virgatus A. Gray var. *asa-grayi* (Crosswh.) Dorn [HL]
Penstemon whippleanus A. Gray [HL]

Plantain
Plantago lanceolata L. [HL] □
Plantago major L. var. *major* [HL]
Plantago patagonica Jacq. var. *patagonica* [HL]

Yellow rattle
Rhinanthus minor L. subsp. *minor* [H]

Speedwell
Veronica americana Schwein. ex Benth. [HL]
Veronica anagallis-aquatica L. [HL]
Veronica peregrina L. var. *xalapensis* (Kunth) H. St. John & F. W. Warren [HL]
Veronica serpyllifolia L. subsp. *humifusa* (Dicks.) Vahl [HL]
Veronica wormskjoldii Roem. & Schult. [HL]

FAMILY POACEAE, GRASSES

Needlegrass, ricegrass (often listed as the genus *Stipa* L.)
Achnatherum hymenoides (Roem. & Schult.) Barkworth [HL]
Achnatherum occidentale (Thurb.) Barkworth subsp. *pubescens* (Vasey) Barkworth [H]
Achnatherum robustum (Vasey) Barkworth [L]
Achnatherum scribneri (Vasey) Barkworth [L]

Wheatgrass
Agropyron cristatum (L.) Gaertn. subsp. *pectinatum* (M. Bieb.) Tzvelev [HL]
Agropyron cristatum (L.) Gaertn. var. *desertorum* (Fisch. ex Link) Dorn [L]

Bentgrass
Agrostis exarata Trin. [L]
Agrostis gigantea Roth [L]
Agrostis scabra Willd. var. *scabra* [HL]
Agrostis stolonifera L. [L]

Meadow foxtail
Alopecurus aequalis Sobol. var. *aequalis* [HL]

Bluestem, beard grass
Andropogon gerardii Vitman [L]
Andropogon hallii Hack. [L]

Threeawn
Aristida adscensionis L. [L]
Aristida arizonica Vasey [L]
Aristida divaricata Humb. & Bonpl. ex Willd. [HL]
Aristida purpurea Nutt. var. *fendleriana* (Steud.) Vasey [H]
Aristida purpurea Nutt. var. *longiseta* (Steud.) Vasey [L]
Aristida purpurea Nutt. var. *nealleyi* (Vasey) Allred [L]

Alpine oat grass
Avenula hookeri (Scribn.) Holub [H]

Sloughgrass
Beckmannia syzigachne (Steud.) Fernald [H]

Pine dropseed
Blepharoneuron tricholepis (Torr.) Nash [HL]

Beardgrass
Bothriochloa laguroides (DC.) Herter subsp. *torreyana* (Steud.) Allred & Gould [L]
Bothriochloa springfieldii (Gould) Parodi [L]

Grama
Bouteloua curtipendula (Michx.) Torr. var. *caespitosa* Gould & Kapadia [L]

Buffalo grass
Bouteloua dactyloides (Nutt.) Columbus [L]
Bouteloua eriopoda (Torr.) Torr. [L]

Blue grama
Bouteloua gracilis (Kunth) Lag. ex Griffiths var. *gracilis* [HL]
Bouteloua hirsuta Lag. var. *hirsuta* [L]

Brome, chess
Bromus anomalus Rupr. ex E. Fourn. [L]
Bromus arvensis L. [L]
Bromus canadensis Michx. [HL]
Bromus inermis Leyss. var. *inermis* [H]
Bromus lanatipes (Shear) Rydb. [HL]
Bromus pumpellianus Scribn. [H]
Bromus racemosus L. [HL]

Cheatgrass
Bromus tectorum L. [HL] □

Reed bent grass
Calamagrostis canadensis (Michx.) P. Beauv. var. *canadensis* [HL]
Calamagrostis purpurascens R. Br. var. *purpurascens* [H]

Sandreed
Calamovilfa longifolia (Hook.) Scribn. var. *longifolia* [L]

Sandbur
Cenchrus longispinus (Hack.) Fernald [L] □

Fingergrass
Chloris verticillata Nutt. [L]

Orchard grass
Dactylis glomerata L. [HL]

Oatgrass
Danthonia parryi Scribn. [HL]

Tufted hair grass
Deschampsia cespitosa (L.) Beauv. var. *cespitosa* [H]

Panic grass
Dichanthelium linearifolium (Scribn.) Gould [L]
Dichanthelium oligosanthes (Schult.) Gould var. *scribnerianum* (Nash) Gould [L]

Crab grass
Digitaria californica (Benth.) Henrard [L]

Barnyard grass
Echinochloa crus-galli (L.) P. Beauv. var. *crus-galli* [L]

Wild rye, bottle-brush grass
Elymus canadensis L. var. *canadensis* [L]
Elymus elymoides (Raf.) Swezey subsp. *brevifolius* (J. G. Sm.) Dorn [HL]
Elymus elymoides (Raf.) Swezey var. *elymoides* [HL]
Elymus glaucus Buckley var. *glaucus* [H]
Elymus lanceolatus (Scribn. & J. G. Sm.) Gould subsp. *lanceolatus* [H]
Elymus repens (L.) Gould [HL] □
Elymus scribneri (Vasey) M. E. Jones [H]
Elymus trachycaulus (Link) Gould ex Shinners subsp. *subsecundus* (Link) Á. Löve & D. Löve [HL]
Elymus trachycaulus (Link) Gould ex Shinners subsp. *trachycaulus* [HL]

Western wheatgrass
Elymus smithii (Rydb.) Gould [HL]

Beardless wildrye
Elymus triticoides Buckley [L]

Lovegrass
Eragrostis cilianensis (All.) Vignolo ex Janch. [L]
Eragrostis curvula (Schrad.) Nees var. *conferta* Stapf [L]
Eragrostis minor Host [L]
Eragrostis pectinacea (Michx.) Nees var. *pectinacea* [L]

Eragrostis pilosa (L.) P. Beauv. var. *perplexa* (L. H. Harv.) S. D. Koch [L]
Eragrostis secundiflora J. Presl subsp. *oxylepis* (Torr.) S. D. Koch [L]
Eragrostis spectabilis (Pursh) Steud. [L]
Eragrostis trichodes (Nutt.) Alph. Wood [L]

Woollygrass
Erioneuron pilosum (Buckley) Nash [L]

Fescue
Festuca arizonica Vasey [HL]
Festuca brachyphylla Schult. & Schult. f. subsp. *coloradensis* Fred. [HL]
Festuca campestris Rydb. [H, rare]
Festuca idahoensis Elmer subsp. *idahoensis* [HL]
Festuca saximontana Rydb. var. *saximontana* [HL]
Festuca thurberi Vasey [H]

Mannagrass
Glyceria grandis S. Watson [L]
Glyceria striata (Lam.) Hitchc. var. *stricta* (Scribn.) Fernald [L]

False alpine or Morton's oatgrass
Helictotrichon mortonianum (Scribn.) Henrard [H]

Needlegrass
Hesperostipa comata (Trin. & Rupr.) Barkworth subsp. *comata* [HL]
Hesperostipa neomexicana (Thurb.) Barkworth [HL]

Northern sweetgrass
Hierochloe hirta (Schrank) Borbás subsp. *arctica* (J. Presl) G. Weim. [HL]

Barley
Hordeum jubatum L. subsp. *intermedium* Bowden [L]
Hordeum pusillum Nutt. [L]
Hordeum vulgare L. [L]

Junegrass
Koeleria macrantha (Ledeb.) Schult. [HL]

Cut grass, white grass
Leersia oryzoides (L.) Sw. [L]

Sprangletop
Leptochloa dubia (Kunth) Nees [L, rare]
Leptochloa fusca (L.) Kunth subsp. *fascicularis* (Lam.) N. Snow [L]

Wolfstail
Lycurus phleoides Kunth [L]
Lycurus setosus (Nutt.) C. Reeder [L]

Muhly
Muhlenbergia arenicola Buckley [L]

Muhlenbergia asperifolia (Nees & Meyen ex Trin.) Parodi [L]
Muhlenbergia cuspidata (Torr. ex Hook.) Rydb. [L]
Muhlenbergia minutissima (Steud.) Swallen [L]
Muhlenbergia montana (Nutt.) Hitchc. [HL]
Muhlenbergia porteri Scribn. ex Beal [L]
Muhlenbergia racemosa (Michx.) Britton, Sterns & Poggenb. [L]
Muhlenbergia repens (J. Presl) Hitchc. [L]
Muhlenbergia torreyi (Kunth) Hitchc. ex Bush [L]
Muhlenbergia wrightii Vasey ex J. M. Coult. [L]

False buffalograss
Munroa squarrosa (Nutt.) Torr. [L]

Green needlegrass, tussock grass
Nassella viridula (Trin.) Barkworth [HL]

Panic grass
Panicum capillare L. [L]
Panicum hallii Vasey var. *hallii* [L]
Panicum obtusum Kunth [L]
Panicum virgatum L. [HL]

Timothy
Phleum alpinum L. var. *alpinum* [H]
Phleum pratense L. var. *pratense* [H]

Reed
Phragmites australis (Cav.) Trin. ex Steud. [L]

Mountain rice grass
Piptatherum micranthum (Trin. & Rupr.) Barkworth [HL]

Galleta grass
Pleuraphis jamesii Torr. [L]

Bluegrass
Poa alpina L. [H]
Poa annua L. [H]
Poa arctica R. Br. subsp. *aperta* (Scribn. & Merr.) Soreng [H]
Poa arida Vasey [L]
Poa bigelovii Vasey & Scribn. [L]
Poa compressa L. [HL]
Poa cusickii Vasey var. *cusickii* [H]
Poa fendleriana (Steud.) Vasey [HL]
Poa leptocoma Trin. [H]
Poa nemoralis L. subsp. *interior* (Rydb.) W. A. Weber [H]
Poa palustris L. [H]
Poa pratensis L. [HL]
Poa reflexa Vasey & Scribn. [L]
Poa secunda J. Presl [H]
Poa tracyi Vasey [H]

Annual beardgrass
Polypogon monspeliensis (L.) Desf. [L]

Bluebunch wheatgrass
Pseudoroegneria spicata (Pursh) Á. Löve subsp. *inermis* (Scribn. & J. G. Sm.) Á. Löve [HL]

Alkali grass
Puccinellia nuttalliana (Schult.) Hitchc. [H]

Tumblegrass
Schedonnardus paniculatus (Nutt.) Trel. [L]

Little bluestem
Schizachyrium scoparium (Michx.) Nash var. *scoparium* [L]

Burrograss
Scleropogon brevifolius Phil. [L]

Foxtail
Setaria leucopila (Scribn. & Merr.) K. Schum. [L]
Setaria pumila (Poir.) Roem. & Schult. subsp. *pumila* [L] ☐
Setaria viridis (L.) P. Beauv. [L] ☐

Indiangrass
Sorghastrum nutans (L.) Nash [L]

Wedge grass
Sphenopholis obtusata (Michx.) Scribn. var. *major* (Torr.) Erdman [L]

Dropseed
Sporobolus airoides (Torr.) Torr. var. *airoides* [L]
Sporobolus compositus (Poir.) Merr. var. *compositus* [L]
Sporobolus cryptandrus (Torr.) A. Gray [HL]
Sporobolus nealleyi Vasey [L]

Tall wheatgrass
Thinopyrum ponticum (Podp.) Barkworth & D. R. Dewey [L]

False mannagrass
Torreyochloa pallida (Torr.) G. L. Church var. *pauciflora* (J. Presl) J. I. Davis [H]

Purpletop
Tridens muticus (Torr.) Nash var. *elongatus* (Buckley) Shinners [L]

Sandgrass
Triplasis purpurea (Walter) Chapman [L]

Yellow oats
Trisetum montanum Vasey var. *montanum* [HL]

Sixweeks grass
Vulpia octoflora (Walter) Rydb. [HL]

FAMILY POLEMONIACEAE

Cheat gily-flower
Aliciella pinnatifida (Nutt. ex A. Gray) J. M. Porter [H]

Mountain trumpet
Collomia linearis Nutt. [HL]

Gilia
Gilia clokeyi H. Mason [L]

Gily-flower
Giliastrum acerosum (A. Gray) Rydb. [L]

Scarlet gilia, sky rocket
Ipomopsis aggregata (Pursh) V. E. Grant subsp. *candida* (Rydb.) V. E. Grant & A. D. Grant [HL]
Ipomopsis aggregata (Pursh) V. E. Grant subsp. *collina* (Greene) Wilken & Allard [HL]
Ipomopsis laxiflora (Coult.) V. E. Grant [HL]
Ipomopsis spicata (Nutt.) V. E. Grant subsp. *spicata* [L]

Desert trumpets
Linanthus pungens (Torr.) J. M. Porter & L. A. Johnson [HL]

Annual phlox
Microsteris gracilis (Douglas ex Hook.) Greene subsp. *humilior* (Greene) H. Mason [HL]

Phlox
Phlox condensata (A. Gray) E. E. Nelson [H]
Phlox longifolia Nutt. subsp. *longifolia* [H]
Phlox pulvinata (Wherry) Cronquist [H]

Jacob's ladder
Polemonium brandegeei (A. Gray) Greene [HL]
Polemonium foliosissimum A. Gray [HL]
Polemonium pulcherrimum Hook. subsp. *delicatum* (Rydb.) Brand [H]

FAMILY POLYGALACEAE

Milkwort
Polygala alba Nutt. var. *alba* [L]

FAMILY POLYGONACEAE

Wild buckwheat
Eriogonum alatum Torr. var. *alatum* [HL]
Eriogonum annuum Nutt. [L]
Eriogonum arcuatum Greene var. *xanthum* (Small) Reveal [H]
Eriogonum cernuum Nutt. [HL]
Eriogonum effusum Nutt. [HL]
Eriogonum jamesii Benth. var. *jamesii* [HL]

Eriogonum lachnogynum Torr. ex Benth. var. *lachnogynum* [HL]
Eriogonum lonchophyllum Torr. & A. Gray var. *fendlerianum* (Benth.) Reveal [L]
Eriogonum tenellum Torr. var. *tenellum* [L]
Eriogonum umbellatum Torr. var. *aureum* (Gand.) Reveal [H]
Eriogonum umbellatum Torr. var. *umbellatum* [H]

Smartweed, knotweed
Polygonum amphibium L. var. *emersum* Michx. [L]
Polygonum amphibium L. var. *stipulaceum* N. Coleman [HL]
Polygonum arenastrum Jord. ex Boreau [HL]
Polygonum bistortoides Pursh [HL]
Polygonum convolvulus L. var. *convolvulus* [HL]
Polygonum lapathifolium L. [L]
Polygonum pensylvanicum L. [L]
Polygonum persicaria L. [L]
Polygonum ramosissimum Michx. var. *ramosissimum* [L]

Dock, sorrel
Rumex acetosella L. [HL]
Rumex altissimus Alph. Wood [L]
Rumex crispus L. [HL] □
Rumex occidentalis S. Watson [L]
Rumex salicifolius Weinm. var. *mexicanus* (Meisn.) C. L. Hitchc. [L]

FAMILY POLYPODIACEAE (SEE CHAPTER 9, FERNS)

FAMILY PONTEDERIACEAE

Mudplantain
Heteranthera limosa (Sw.) Willd. [L]

FAMILY PORTULACACEAE

Spring beauty
Claytonia rosea Rydb. [H]

Bitterroot
Lewisia nevadensis (A. Gray) B. L. Rob. [H]

Candy flower or toad lily
Montia chamissoi (Ledeb. ex Spreng.) Greene [HL]

Fameflower
Phemeranthus confertiflorus (Greene) Hershk. [L]

Purslane
Portulaca halimoides L. [L, rare]
Portulaca oleracea L. [L] □

FAMILY POTAMOGETONACEAE

Pondweed
Potamogeton diversifolius Raf. [L, rare]
Potamogeton foliosus Raf. [L]
Potamogeton pusillus L. var. *pusillus* [HL]

Sago pondweed, false pondweed
Stuckenia pectinata (L.) Börner [HL]

FAMILY PRIMULACEAE

Rock jasmine
Androsace chamaejasme Host subsp. *lehmanniana* Spreng. [H]
Androsace septentrionalis L. [H]
Androsace septentrionalis L. subsp. *glandulosa* (Wooton & Standl.) G. T. Robbins [L]
Androsace septentrionalis L. subsp. *puberulenta* (Rydb.) G. T. Robbins [HL]
Androsace septentrionalis L. subsp. *subulifera* (A. Gray) G. T. Robbins [HL]

Shooting star
Dodecatheon pulchellum (Raf.) Merr. var. *pulchellum* [HL]

Loosestrife
Lysimachia ciliata L. [L]
Lysimachia thyrsiflora L. [L, extirpated]

Primrose
Primula angustifolia Torr. [H]
Primula parryi A. Gray [H]

FAMILY PTERIDACEAE (SEE CHAPTER 9, FERNS)

FAMILY RANUNCULACEAE

Monkshood
Aconitum columbianum Nutt. var. *columbianum* [HL]

Baneberry
Actaea rubra (Aiton) Willd. [HL]

Windflower
Anemone canadensis L. [HL]
Anemone cylindrica A. Gray [L]

Pasqueflower, prairie smoke
Anemone multifida Poir. [H]
Anemone virginiana L. var. *virginiana* [H, rare]

Prairie crocus
Anemone patens L. var. *multifida* Pritz. [HL]

Columbine
Aquilegia coerulea E. James var. *coerulea* [H]
Aquilegia elegantula Greene [H]

Marsh marigold
Caltha leptosepala DC. [H]

Virgin's bower, clematis
Clematis columbiana (Nutt.) Torr. & A. Gray var. *columbiana* [HL]
Clematis hirsutissima Pursh [HL]
Clematis ligusticifolia Nutt. [HL]

Larkspur
Delphinium alpestre Rydb. [HL]
Delphinium barbeyi (Huth) Huth [H]
Delphinium ramosum Rydb. [HL]
Delphinium virescens Nutt. [HL]
Delphinium wootonii Rydb. [HL]

Buttercup
Ranunculus cardiophyllus Hook. [HL]
Ranunculus cymbalaria Pursh [HL]
Ranunculus inamoenus Greene var. *inamoenus* [HL]
Ranunculus macauleyi Gray [HL]
Ranunculus macounii Britton [H]
Ranunculus sceleratus L. var. *sceleratus* [L]
Ranunculus trichophyllus Chaix var. *trichophyllus* [H]

Meadow-rue
Thalictrum dasycarpum Fisch. & Avé-Lall. [L]
Thalictrum fendleri Engelm. ex A. Gray var. *fendleri* [HL]

FAMILY RHAMNACEAE (SEE CHAPTER 10, WOODY FLOWERING PLANTS)

FAMILY ROSACEAE (SEE ALSO CHAPTER 10, WOODY FLOWERING PLANTS)

Mountain avens
Dryas octopetala L. subsp. *hookeriana* (Juz.) Breitung [H]

Strawberry
Fragaria vesca L. [L]
Fragaria virginiana Mill. [HL]

Avens
Geum aleppicum Jacq. [HL]
Geum macrophyllum Willd. var. *perincisum* (Rydb.) Raup [HL]
Geum rivale L. [L]

Cinquefoil
Potentilla arguta Pursh subsp. *arguta* [HL]
Potentilla diversifolia Lehm. var. *diversifolia* [H]

Potentilla gracilis Douglas ex Hook. var. *brunnescens* (Rydb.) C. L. Hitchc. [HL]
Potentilla hippiana Lehm. var. *hippiana* [HL]
Potentilla norvegica L. subsp. *monspeliensis* (L.) Asch. & Graebn. [H]
Potentilla pensylvanica L. var. *pensylvanica* [HL]
Potentilla pulcherrima Lehm. [HL]
Potentilla subviscosa Greene [L]

Shrubby cinquefoil, golden hardhack
Potentilla fruticosa L. [HL]

Rose
Rosa acicularis Lindl. subsp. *sayi* (Schwein.) W. H. Lewis [L]
Rosa arkansana Porter [L]
Rosa manca Greene [HL]

Raspberry, blackberry
Rubus deliciosus Torr. [L]
Rubus idaeus L. subsp. *strigosus* (Michx.) Focke [HL]

Burnet
Sanguisorba minor Scop. subsp. *muricata* Briq. [HL]

Creeping glow-wort
Sibbaldia procumbens L. [H]

FAMILY RUBIACEAE

Bedstraw
Galium aparine L. var. *aparine* [L]
Galium boreale L. [HL]
Galium triflorum Michx. [HL]

FAMILY SALICACEAE (SEE CHAPTER 10, WOODY FLOWERING PLANTS)

FAMILY SANTALACEAE

Bastard toadflax
Comandra umbellata (L.) Nutt. subsp. *pallida* (A. DC.) Piehl [HL]

FAMILY SAPINDACEAE (SEE CHAPTER 10, WOODY FLOWERING PLANTS)

FAMILY SARCOBATACEAE

Black greasewood
Sarcobatus vermiculatus (Hook.) Torr. [HL]

FAMILY SAXIFRAGACEAE

Alumroot
Heuchera parviflora Nutt. ex Torr. & A. Gray [HL]
Heuchera rubescens Torr. var. *versicolor* (Greene) M. G. Stewart [L, extirpated]

Saxifrage
Saxifraga bronchialis L. subsp. *austromontana* (Wiegand) Piper [L]
Saxifraga flagellaris Willd. ex Sternb. subsp. *crandallii* (Gand.) Hultén [H]
Saxifraga odontoloma Piper [H]
Saxifraga rhomboidea Greene [HL]

FAMILY SCROPHULARIACEAE

Figwort
Scrophularia lanceolata Pursh [H]

Mullein
Verbascum thapsus L. [HL] ☐

FAMILY SELAGINELLACEAE (SEE CHAPTER 9, FERNS)

FAMILY SOLANACEAE

Five eyes
Chamaesaracha coniodes (Moric.) Britton [L]
Chamaesaracha coronopus (Dunal) A. Gray [L]

Wolfberry
Lycium pallidum Miers [L]

Groundcherry
Physalis hederifolia A. Gray var. *comata* (Rydb.) Waterf. [L]
Physalis hederifolia A. Gray var. *fendleri* (A. Gray) Cronquist [H]
Physalis hispida (Waterf.) Cronquist [L]
Physalis subulata Rydb. var. *neomexicana* (Rydb.) Waterf. ex Kartesz & Gandhi [L]
Physalis virginiana Mill. [L]

Chinese lantern
Quincula lobata (Torr.) Raf. [HL]

Nightshade
Solanum elaeagnifolium Cav. [L] ☐
Solanum jamesii Torr. [L]
Solanum ptycanthum Dunal [L] ☐
Solanum rostratum Dunal [L]
Solanum triflorum Nutt. [HL]

FAMILY TAMARICACEAE

Tamarisk, salt cedar
Tamarix chinensis Lour. [HL] ☐

FAMILY TYPHACEAE

Cattail
Typha domingensis Pers. [L]
Typha latifolia L. [L]

FAMILY ULMACEAE (SEE CHAPTER 10, WOODY FLOWERING PLANTS)

FAMILY URTICACEAE

Pellitory
Parietaria pensylvanica Muhl. ex Willd. [L]

Nettle
Urtica dioica L. var. *procera* (Muhl. ex Willd.) Wedd. [HL]

FAMILY VALERIANACEAE

Sharpleaf valerian
Valeriana acutiloba Rydb. var. *acutiloba* [H]

FAMILY VERBENACEAE

Mock vervain
Glandularia bipinnatifida (Nutt.) Nutt. var. *bipinnatifida* [HL]
Glandularia bipinnatifida (Nutt.) Nutt. var. *ciliata* (Benth.) B. L. Turner [HL]

Fogfruit
Phyla cuneifolia (Torr.) Greene [L]
Phyla lanceolata (Michx.) Greene [L]

Vervain
Verbena bracteata Lag. & Rodr. [HL]
Verbena hastata L. var. *scabra* Moldenke [L]
Verbena macdougalii A. Heller [L]
Verbena plicata Greene [L]
Verbena stricta Vent. [L]

FAMILY VIOLACEAE

Green violet
Hybanthus verticillatus (Ortega) Baill. [L]

Violet
Viola adunca Sm. [H]
Viola canadensis L. [HL]
Viola labradorica Schrank [H]
Viola macloskeyi F. E. Lloyd subsp. *pallens* (Banks ex DC.) C. L. Hitchc. [L]
Viola nephrophylla Greene [HL]
Viola nuttallii Pursh [HL]
Viola pedatifida G. Don [HL, rare]
Viola praemorsa Douglas ex Lindl. [L]

FAMILY VISCACEAE

Dwarf mistletoe

Arceuthobium cyanocarpum (A. Nelson ex Rydb.) A. Nelson [H]
Arceuthobium vaginatum (Willd.) J. Presl subsp. *cryptopodum* (Engelm.) Hawksw. & Wiens [L] (Fig. 91)

FAMILY VITACEAE

Woodbine

Parthenocissus vitacea (Knerr) Hitchc. [L]

Grape

Vitis acerifolia Raf. [L]
Vitis riparia Michx. [L]

FAMILY ZANNICHELLIACEAE

Horned pondweed

Zannichellia palustris L. [HL]

FAMILY ZYGOPHYLLACEAE

Caltrop

Kallstroemia parviflora Norton [L]

Puncture vine

Tribulus terrestris L. [HL] □

Fig. 91: Dwarf mistletoe (*Arceuthobium* M. Bieb. sp.), a parasitic plant, on a conifer branch.

CHAPTER 13

LOCAL FAUNA

DESPITE the growing numbers of people in this region, there remains ample wild land to support the survival of many species of animals long absent from more urban areas.

Surprisingly, there is no current state guide to these animals and we are not even close to an authoritative checklist for Huerfano or Las Animas counties. The following list, from various sources, includes comments on the more readily seen animals, plus comments on rumors of those no longer encountered here. Our knowledge of many small and less conspicuous animals is uneven, as many are rarely seen. They may be present but are undocumented. I have tried to err on the side of including possible residents. The National Museum of Natural History (MNH) website allows searches on all North American mammals by region. It also allows the downloading of species pages to build your personal field guide.

RESPECT THEIR SPACE

When observing wild animals, keep your distance. Do not attempt to feed them. Animals habituated to human presence may not only become dangerously unpredictable but usually have their lives cut short.

Sadly, the most likely encounter a visitor will have with many of our species is as road kill. Nothing is as efficient as the road system at harvesting animals. Deer and elk cause the most damage to cars as they pay the ultimate sacrifice. Large animals and many smaller species seem especially confused by headlights. To avoid making your contribution to this carnage, be alert while driving, especially at dawn and dusk. (See Knutson's field guide [1987].)

Various regional and national field guides are valuable in identification of animal species. Specialists make definitive determinations of mammals, especially smaller ones, from skull and dentition evidence. For reptiles, scale counts around the head are important because colors and patterns can vary with age and ecology.

MAMMALS

OPOSSUMS, FAMILY DIDELPHIDAE

Virginia opossum (*Didelphis virginiana*)

Opossums are recent migrants to Colorado. Females of these marsupials carry their young in a pouch. These light gray animals reach 2 feet long overall and weigh more than 7 pounds. They have a pointed face and naked tail. Considered the ultimate generalists, opossums are omnivores, have opposable toes that aid climbing, and will den in nearly any sheltered spot. They are noted for their ability to play dead, but they have many sharp teeth, so approach them cautiously.

SHREWS, FAMILY SORICIDAE

Shrews are smaller than mice and usually short tailed. Their fur is dark gray. They move quickly, have a high metabolism, and have a heart rate of 600 to 1200 beats per minute. To stoke this "fire," they are carnivores, eating small animals, eggs, larvae, insects, or carrion. They are also noted for their strong odor, and other predators mostly leave them alone. Shrews range from Great Plains elevation to more than 10,000 feet.

In this region are the desert shrew (*Notiosorex crawfordii*) in both counties, the masked shrew (*Sorex cinereus*) in Huerfano County only, the montane shrew (*S. monticolus*) in both counties, and the water shrew (*S. palustris*) in both counties.

MOLES, FAMILY TALPIDAE

Eastern mole (*Scalopus aquaticus*)

These are fossorial (underground-dwelling) mammals. So far they are only found in the eastern plains of Colorado. Characteristically, they excavate the annoying raised tunnels on manicured lawns. In nature, this tunneling makes them effective soil builders. For food they favor worms and insects.

BATS, FAMILY VESPERTILIONIDAE

Compared with other mammal groups, bats are well documented for the state. Locally, the San Luis Valley is well surveyed scientifically, but Huerfano and Las Animas counties appear to be little known. The species we have are insectivorous. They catch flying insects in the dark using echolocation, which explains their outsized ears.

As they are vulnerable to predation and nearly blind during the day, bats need protected daytime roost sites such as tree holes or branches, or crevices in rock faces, chimneys, in old buildings, behind shutters, or under roof eaves.

Occasionally, bats get into cabins, but please refrain from killing them. Open the doors and windows and they will leave a lit cabin for the dark. If you exclude bats that have been living in chimneys or crevices for many years, erect a bat house near the cabin to provide a daytime roosting site.

Bat-transmitted rabies, the only potential problem, is rare. Only about 0.5% of bats ever get rabies and less than 1% of those bite anything. Nearly all bats that get rabies are paralyzed, cannot fly, and die quickly. There have been very few such human bites in Colorado and no human fatalities. The value of bats to humans and to the ecology is enormous. They eat many thousands of insects, many of which are agricultural pests. One bat may kill as many as 3000 insects per night, including mosquitoes.

Big brown bat (*Eptesicus fuscus*)

A common North American bat, they are found in all habitats.

Big free-tailed bat (*Nyctinomops macrotis*)

This is the largest Colorado bat, but there are only a few scattered records.

Brazilian free-tailed bat (*Tadarida brasiliensis*)

This bat is found in pinyon-juniper, semi-desert, and grassland habitats.

Hoary bat (*Lasiurus cinereus*)

They forage around irrigation canals and may be seen in ponderosa pine forests.

Long-eared myotis (*Myotis evotis*)

Found from 5000- to 9800-foot elevation, they have been recorded from the La Veta Pass area. Among local bats, their ears are the longest. Black ear membranes contrast with yellowish-brown fur. They are always found in coniferous forests, where they roost in trees behind bark or in sheds, cabins, caves, or old mines.

Little brown bat (*Myotis lucifugus*)

Once recorded from the La Veta Pass area, they have buff fur and large but pale ears. During the day they roost in tree hollows, beneath bark, under buildings, rock crevices, or under bridges. Hibernation is in caves and old mine shafts, and they are not yet known from the Spanish Peaks neighborhood.

ARMADILLOS, FAMILY DASYPODIDAE

Nine-banded armadillo (*Dasypus novemcinctus*)

This southern animal is expanding its range and has now reached eastern Colorado counties. Like opossums, they are generalist omnivores and live in a variety of temporary protected sites.

PIKAS, FAMILY OCHOTONIDAE

Pika, cony (*Ochotona princeps*)

This small "rock rabbit," a distant relative of the cottontail, is found above timberline on talus slopes or rocky areas. They gather hay, placing vegetative materials in stacks to dry before moving it into their dens for the winter. They can be identified by their high-pitched whistle, often before they are seen.

RABBITS, FAMILY LEPORIDAE

Two genera are present: cottontails and hares or jackrabbits.

Cottontails, the smaller animals, are as common at lower elevations here as elsewhere in the United States. They are occasionally seen up to at least 10,000-foot elevation. They eat vegetation.

Eastern cottontail (*Sylvilagus floridanus*)

Known from Las Animas County and possibly present in Huerfano County.

Mountain cottontail, Nuttall's cottontail (*Sylvilagus nuttallii*)

This species is found in both counties and extends northward to the northwest section of the state. Usually found at the pinyon-juniper elevation, they occur as high as 9000 to 10,000 feet.

Desert cottontail (*Sylvilagus audubonii*)

Lives mostly in brushy country in both counties.

Jackrabbits and hares (*Lepus townsendii, L. americanus*)

Jackrabbits and hares are much larger than cottontails, with large ears and enlarged rear legs capable of fast bounding. They may reach 2 feet in length and weigh 6 to 9 pounds. They favor mountain parks and lower-elevation sagebrush country, where they eat herbs in the growing season and bark in the winter. Their fur is white in winter. Ours are the white-tailed jackrabbit (*Lepus townsendii*) and the snowshoe hare (*L. americanus*).

SQUIRRELS, CHIPMUNKS, FAMILY SCIURIDAE

Chipmunks are lively and familiar animals. They often beg for food that they store in cheek pouches. Although not deep hibernators, they overwinter by sleeping for several days at a time. Their burrows are usually beneath a rock or stump. As with other squirrels, they will bite occasionally, so refrain from training them to take food from your hand. Their regular diet is berries, seeds, flowers, and insects.

Least chipmunk (*Tamias minimus*)

The most widespread in the state, especially central and western Colorado.

Colorado chipmunk (*Tamias quadrivittatus*)

Most common in southern Colorado, they are recognized by their striped back. They are documented for Las Animas County and likely occur in Huerfano County. Total length is up to 9 inches, and they weigh about 2 ounces.

Wood rats and pack rats are plant eaters.

Bushy-tailed woodrat (*Neotoma cinerea*)

They prefer montane to subalpine forests, high-elevation exposed places with numerous boulders, and old mining camps. They will occasionally live in abandoned buildings. They are also known from pinyon-juniper woodlands and semi-desert scrublands. Their coat color is buff to gray.

Mexican woodrat (*Neotoma mexicana*)

These animals prefer rim rock and other rocky areas for making nests and burrows. Their burrows are filled with large quantities of materials, and they eat fruits and seeds. Including the 5-inch tail, they may reach 12 inches long.

Southern Plains woodrat (*Neotoma micropus*)

A denizen of semi-arid grasslands and shrublands of the eastern plains counties, they are probably not in this guidebook's territory. They are known from eastern Las Animas County. Their fur is gray, and they eat plant material.

White-throated woodrat (*Neotoma albigula*)

Present in both counties, their habits are little known.

Eastern woodrat or packrat (*Neotoma floridana*)

A beautiful gray animal found on the plains or riparian woodlands of counties north and east of here. Although not officially documented from these two counties, some claim to have observed them. They pick up curious, often shiny objects that they use to line their dens, spaces sometimes in use for many generations. Packrat dens thousands of years old have yielded important information at archeological sites.

Yellow-bellied marmot (*Marmota flaviventris*)

This groundhog-sized animal occurs from the lower foothills to alpine meadows in both counties. It reaches about 26 inches long and weighs 11 pounds. Marmots make large burrows in deep soil of open areas, usually between boulders. They are often seen sunning themselves but will dive for their burrow when approached by hikers. They are true hibernators that spend the growing season napping, sunning themselves, grooming, and foraging for their diet of flower stalks and leaves.

Ground squirrels are about the size of tree squirrels; these animals have a striped or spotted fur pattern on their backs.

Spotted ground squirrel (*Spermophilus spilosoma*)

An eastern Plains species, it is known from Las Animas but not from Huerfano County. They are grayish brown with dorsal buffy spots.

Thirteen-lined ground squirrel (*Spermophilus tridecemlineatus*)

Restricted to grasslands, they are most numerous on heavily grazed areas including prairie dog towns. They have alternating rows of white stripes and white spots. As omnivores, they eat plant material, some insects, and some dead mammals.

Golden-mantled ground squirrel (*Spermophilus lateralis*)

Looking like a giant chipmunk, they are called "mini-bears" at Philmont Scout Ranch to the south of our territory. They will get into packs and chew things besides your food if you camp at established camps. When camping, keep your belongings as secure as you would when bears are in the area. Not documented from Las Animas County but they must be there.

Rock squirrel (*Spermophilus variegatus*)

Mostly in western mesas and canyons but also reported locally.

Black-tailed prairie dog (*Cynomys ludovicianus*)

According to a recent census, more than 631,000 acres in Colorado provide habitat for prairie dogs, a much greater range than recently thought. They are in Las Animas County but are not reported yet from Huerfano County. Their fur is dun colored, and they have the habit of standing upright on their burrow mounds. Although many people find them interesting to watch, ranchers usually take a dim view of prairie dogs because of their capacity for destroying rangeland and for making holes in which large animals can break their legs.

Abert's squirrel (*Sciurus aberti*)

This species is not documented from Las Animas County. It is more common in central and western Colorado. Among squirrel species, they are recognized by their conspicuous ear tufts. Abert's squirrels favor ponderosa pine stands for habitat.

Red squirrel, chickaree, pine squirrel (*Tamiasciurus hudsonicus*)

Small, noisy, reddish squirrels that live in high-elevation conifer forests. Their persistent alarm chattering announces your presence into their territory. With years of effort, pine squirrel populations will leave huge middens of piled cone scales that are seen along high-elevation forest trails throughout the region.

POCKET GOPHERS, FAMILY GEOMYIDAE

Ground dwelling and seldom seen, pocket gophers have yellow-faced front teeth. They have fur-lined cheek pouches where they temporarily store their gathered seeds. Their burrows may be as long as 200 yards, and they make no mounds except under snow.

Valley pocket gopher (*Thomomys bottae*)

This species is reddish brown with a blackish chin and reddish belly. It is found here and in southern and western valleys.

Northern pocket gopher (*Thomomys talpoides*)

This species is dark colored with a whitish chin and belly. It is known from the northern and western fringes of this guide's range, but mostly from the mountains and northwestern part of the state (Fig. 92).

Fig. 92: A soil cast as it appears after snow melt. During the winter, pocket gophers form soil casts from burrows at the soil-snow boundary.

POCKET MICE, FAMILY HETEROMYIDAE

The two genera *Perognathus* and *Chaetodipus* are long-tailed mice that are probably represented here by the following species. Pocket mice also have cheek pouches and are seed eaters. Along roads at night they are the most likely small mammals to freeze in headlights. They favor shortgrass prairie or other grassy habitats.

Hispid pocket mouse (*Chaetodipus hispidus*)

Reported from both counties from grassy sites, it is the largest species in Colorado. Its fur, harsh to the touch, is reddish buff on the back with paler coloration on the sides.

Olive-backed pocket mouse (*Perognathus fasciatus*)

From Huerfano County only, it is olivaceous in color and is the darkest of Colorado species.

Silky pocket mouse (*Perognathus flavus*)

In Las Animas County and possibly Huerfano County. The smallest species, its coat is usually pale buff.

BEAVER, FAMILY CASTORIDAE

Beaver (*Castor canadensis*)

Many streams in the high country are virtually terraced as one beaver dam follows another. Ecologists refer to beavers as "keystone species" because of their capacity to change the environment profoundly in a way that creates habitat for many other animals. The aspens and willows they cut down will resprout, producing thicket-like regrowth that is good for the watershed.

Beavers are large, up to 3 feet long and weighing up to 55 pounds. Their broad, flat tail is a key feature. Their perpetually growing teeth need regular use. A beaver needs about 30 minutes to fell a 5-inch-diameter aspen. They eat the tender twigs, leaves, and bark but not the inner wood.

Although their dams and lodges may look like a flimsy pile of sticks, they are tightly interwoven, mud packed, and very difficult to remove or disturb. In the center of the pond, free of predators, they build a stick lodge accessed from underwater. Along river shores, lodges are built along the bank. Beavers themselves can be seen best around dusk or dawn if you are very quiet in the vicinity of their lakes. When they detect you, a tail slap is a dive signal indicating your last view of them.

In the age of the historic fur-trapping mountain men, beavers were even more common throughout the West. Look for their dams on the Cuchara River on the way to Blue Lake, and also just over La Veta Pass on the north side of Highway 160.

NATIVE RATS, MICE, WOODRATS, VOLES, FAMILY CRICETIDAE

This family is a large and diverse miscellany of interesting small and large mammals.

Field Mice (*Peromyscus*)

The following five species of this genus sometimes make it into human dwellings and are often mistaken for house mice. However, they are less of a nuisance than house mice and are much more easily trapped. All are 6 to 8 inches long, including the tail, and have a buff coat color with white belly. Their faces have large, dark eyes. They eat seeds, leaves, or carrion.

The deer mouse, *Peromyscus maniculatus*, is found statewide in all habitats except wetlands. The brush mouse, *P. boylii*, lives in riparian cottonwood areas, oak thickets, and tamarisk bottoms. The pinyon mouse, *P. truei*, lives in the pinyon-juniper community. The northern rock mouse, *P. nacutus*, is found along mountain front foothills to pinyon-juniper, cliffs, and canyons. The white-footed mouse, *P. leucopus*, is the most common field mouse in the United States. It has varied habitat preferences that include anything from weedy riparian woods to rocky canyon country.

Western harvest mouse (*Reithrodontomys megalotis*)

Smaller than deer mice, this species also has a diverse ecology. They are seed eaters.

Voles or meadow mice (species of the genera *Clethrionomys, Microtus,* and *Phenacomys*)

Difficult to distinguish, voles have pudgy bodies, small eyes, blunt snouts, and short, inconspicuous ears. All are brown to gray and usually have a short tail that is not hairy. Their meadow runways are found beneath a roof of thatch. They eat plant material. Ours are:

Southern red-backed vole (*Clethrionomys gapperi*)

This species favors well-developed conifer forests, especially lodgepole pines. Reported from Huerfano County, they may possibly occur in Las Animas County.

Long-tailed vole (*Microtus longicaudus*)

Known from both counties, they favor streamside meadows from the Great Plains to above timberline.

Prairie vole (*Microtus ochrogaster*)

Reported from Huerfano County, they are most common near streams and irrigated lands.

Meadow vole (*Microtus pennsylvanicus*)

Known from both counties, they favor marshy meadows.

Heather vole (*Phenacomys intermedius*)

Found in both counties at all elevations in forested mountains.

Grasshopper mouse (*Onychomys leucogaster*)

Coyote colored, they are ferocious carnivores with dagger-like teeth. They eat beetles, grasshoppers, other insects, and spiders.

Ord's kangaroo rat (*Dipodomys ordii*)

These seed eaters are at home at lower elevations in the state, including these counties. To escape predation, their outsized rear legs can propel them into high jumps for at least as far as 6 feet. When airborne, their tufted tail acts as a rudder. The animal is about 11 inches long.

Hispid cotton rat (*Sigmodon hispidus*)

A recent arrival from the southern United States via the Arkansas Valley to Pueblo. Not recorded here yet but in northern and eastern neighboring counties. They are about 11 inches long and have blackish-brown fur. The ears are obscured by fur and the tail is mostly naked. Unlike the Norway rat, they stay away from buildings or dumps. They make runways through vegetation and eat plants, eggs, or nestlings.

Muskrat (*Ondatra zibethicus*)

As the largest and most conspicuous species in this family, muskrats may reach more than 20 inches long and weigh nearly 2 pounds. They eat mostly plant material as well as small aquatic animals or carrion. Muskrats live in bank burrows, or usually grassy, conical lodges in ponds and marshes where they often coexist with beavers. They have a laterally flattened, scaly tail.

OLD WORLD RATS AND MICE, FAMILY MURIDAE

These non-natives are the most likely to be found in and around town buildings. Their habits and disease-carrying potential give the native rodents a bad name. They are distinguished from native rodents by their uniformly gray or brown coats and naked, scaly tails. Smarter than the more naive native rodents, they adapt quickly around humans. If you trap one, you probably can't trap the rest of the family by the same method. Poison baits are most effective.

House mouse (*Mus musculus*)

The size of deer mice, but the house mouse is all gray instead of having a white belly.

Norway rat (*Rattus norvegicus*)

Their overall length is 14 to 18 inches. They were carried west to Denver in the 1880s. Today they are common in cities and dumps of the eastern Great Plains–mountain transition in Colorado. They can be bold and aggressive and should be eliminated quickly when seen.

PORCUPINES, FAMILY ERETHIZONTIDAE

Porcupine (*Erethizon dorsatum*)

A cat-sized mammal covered densely with sharp quills. The slow-moving porcupines are woodland animals that eat herbs and bark. Contrary to folk tales, they cannot throw their quills. However, the quills are backward-barbed and work their way deeper into skin, so keep your distance and keep children and pets well clear.

DOGS, FAMILY CANIDAE

Coyote (*Canis latrans*)

The "song dog" is one of the most widespread large animal species in North America. Coyotes are everywhere, especially at lower elevations. In this area, a buff-colored, wild-looking dog with a bushy tail is never a wolf. Coyote populations are expanding despite their being regarded as varmints by ranchers, who use various "coyote getters" to remove them as a perceived threat to young calves.

Our two fox species are up to 3 feet long and weigh 9 to 11 pounds. They eat rodents, rabbits, and birds. Being mostly nocturnal, you may encounter them around dawn or dusk.

Gray fox (*Urocyon cinereoargenteus*)

The tail is black tipped and the ears and feet may be red. They are most often found in brushy areas in the foothills.

Red fox (*Vulpes vulpes*)

The fur is red above, white below, and the tail is white tipped. Red foxes favor riparian woodlands, wetlands on plains, and forest edge.

BEARS, FAMILY URSIDAE

Black bear (*Ursus americanus*) (see also Chapter 6, The Life of Large Charismatic Predators)

Black bears range widely in western North America and in the Appalachian east. Approximately 12,000 are believed to roam in Colorado in all rural areas except the northeastern plains counties. They may be black or occur in various shades of lighter brown (cinnamon to blond). Younger bears between 150 and 300 pounds are most commonly seen, but older animals range in weight up to more than 500 pounds. The bears have a convex, elongated face that may be lighter colored than their bodies. As "generalist" omnivores, they will feed opportunistically on anything from dead carcasses to berries, fruits, ants, bees, and human food and garbage when they can find it. They are about 75% to 90% vegetarian and require about 25,000 calories a day before bedding down for the winter. Although not true hibernators, they retire to winter dens and become very lethargic, punctuated by night or day "activity periods."

RACCOON, FAMILY PROCYONIDAE

Raccoon (*Procyon lotor*)

Easily identified by their black-masked faces and ringed tails, raccoons reach 2 to 3 feet long and weigh 8 to 22 pounds. They are found statewide at intermediate elevations. Like black bears, they adapt easily to humans and eat most anything. On occasion, they have gotten into cabins where they leave a general mess. In nature they prefer to eat near water, as they have the habit of rinsing their food. Car encounters cause the greatest number of deaths.

WEASELS, MARTENS, SKUNKS, BADGERS, OTTERS, FAMILY MUSTELIDAE

A diverse family of active, skilled hunters. Rarely seen, with the exception of skunks.

Pine marten (*Martes americana*)

Martens are found across the northern conifer forests and in the Rockies south to Colorado. They can get up to 2 feet long and are brown with a white belly. You are unlikely to see one, but they are distinguished by their sinuous bounding instead of running. They are weasel-like but broader and more robust. Their long, bushy tail is about one third of their total length. Coat color ranges from buff to dark brown and is distinguished by a vertical black line above the inner corner of each eye. They are climbers that pursue squirrels in dense conifer forests. They eat small animals and eggs, some carrion, and fruit and nuts.

Long-tailed weasel (*Mustela frenata*)

This largest of weasels is widespread in the United States. They are shy and solitary hunters with brown coats that become white in winter. Occasionally in this area you might see one crossing the road. Their agile and slender bodies are up to 18 inches long and seem to bound sinuously when running. Weasels are very alert hunters, preying on anything from rabbits to birds, small rodents, snakes, or turtles, as well as eating fruits and berries. They live in a variety of places including hollow logs, rock piles, and under buildings.

Ermine, short-tailed weasel (*Mustela erminea*)

About 8 to 10 inches long, they become white in winter like the long-tailed weasel. Found mostly in the mountains, their diet is small animals.

Mink (*Mustela vison*)

Mink are found statewide, often near beaver ponds or other still water. They are semi-aquatic, are about 20 to 24 inches long, and weigh about 35 pounds. They have dark brown fur, a black-tipped tail, and partially webbed toes. They live in burrows excavated by others, such as muskrat dens, or in tree hollows. As predators, they eat muskrats, nestling waterfowl, frogs, crayfish, insects, or young beavers. Their leading cause of death is combat with other mink.

Badger (*Taxidea taxus*)

They are the largest weasels, reaching 28 to 32 inches, have a 4- to 6-inch-long tail, and weigh 11 to 22 pounds. They are known in Colorado as animals of loamy soils of the eastern plains, the shrub country of mountain or western valleys, and occasionally above timberline. Badgers are fierce, ill-tempered animals that eat burrowing rodents, ground birds, rabbits, and insects. Habitat loss and cars kill many of them. Badgers are excellent diggers in pursuit of prey and when making their extensive burrow systems.

Skunks (*Conepatus, Mephitus,* and *Spirogale*)

Skunks are distinctive. As is true of many animals that have poison or other chemical defenses, skunks are strikingly marked. The different species have specific white marking patterns on a black coat. As a group, their presence can be detected, often from a distance, by their characteristic odor. However, the odor may also indicate the presence of bears or other animals that have gotten into an altercation with a skunk. If you see one, keep your distance. They signal a spray event by pointing their rear at you and raising their tail.

The hog-nosed skunk (*Conepatus mesoleucus*) is possibly in Las Animas County. The striped skunk (*Mephitis mephitis*) occurs in all Colorado counties. Spotted skunks (*Spirogale gracilis*) are known from Huerfano County but probably occur in both counties.

River otter (*Lutra canadensis*)

This, the largest of the weasel family, may reach a length of more than 42 inches. Its coat is a rich brown. Their sinuous bodies and webbed toes show adaptation to living around water. Otters often live in abandoned beaver bank lodges and they eat young beavers, muskrats, and crayfish. Never abundant in Colorado, they were extirpated in the early 1900s. The Colorado Division of Wildlife reintroduced otters in some rivers in the 1970s. They likely occur in both counties.

CATS, FAMILY FELIDAE

Mountain lion, puma, cougar (*Puma* [or *Felis*] *concolor*) (see also Chapter 6, The Life of Large Charismatic Predators)

This largest of American cats is becoming more common in this area. Being reclusive, they are seen only occasionally in mountain valleys and at higher elevations. Tracks have been reported throughout the mountain areas. You are unlikely to see one unless in the presence of an experienced guide. Their distinctive cry, like a woman's scream, can be heard for miles.

Once distributed throughout the North American continent, they are mostly restricted to the western deserts and mountains, from sea level to 10,000 feet. They are spotted occasionally in the east, including Illinois. Some eastern sightings are the Florida panther subspecies, as a small number still persist in the Everglades.

Large western male cougars may reach 180 pounds, while the smaller female is seldom more than 130 pounds. Their body can reach a length of 6 feet, not including a 32-inch-long black-tipped tail. Their home range varies from 25 to 300 square miles. This plus their stealth and nocturnal lifestyle prevents making an accurate census. However, biologists agree that they are increasing throughout central and southern Colorado.

For prey, cougars are considered deer specialists, but some will attack and eat elk. After a kill, they will gorge themselves, then hide the carcass and revisit it over a period of days. As generalist predators, they will also kill smaller animals down to mice and insects. Where people have moved into their habitat, they will gradually habituate to people's presence and will eat pet food and occasionally pets.

Encountering one is very improbable. If that happens, stay calm and speak firmly to it. Back away slowly. If it appears aggressive, throw stones or branches. Do not turn your back, crouch, or run. Fight back if attacked.

Lynx (*Felis lynx*)

Lynxes are about 3 feet long, are bobtailed, and have conspicuous ear tufts and a solid black tail tip. They prefer dense subalpine forests, avalanche chutes, and dense, willow-covered stream valleys. As carnivores, they are stalkers that favor snowshoe hares, which they usually catch in a single bound. Canadian lynxes have been introduced into southern Colorado. They are reported from Las Animas County and possibly Huerfano County. There are only about 18 confirmed native records in Colorado since 1878 and 11 sets of tracks since 1979. If you see one around here, you will be the first in a very long time.

Bobcat (*Felis rufus*)

In southern Colorado, bobcats are large, bobtailed, and difficult to distinguish from the usually larger lynx. They look a little like cougar cubs, except that cougars of any size always have long tails. Their range is statewide in Colorado, although they tend to avoid open prairies as well as heavy subalpine timber and wetlands.

As with other cats, they are not commonly seen. Bobcats are stealth hunters, predators that only rarely need long chases to catch prey. They prefer to eat rabbits but will eat other small mammals and birds when necessary. They are active throughout the year.

DEER, ELK, FAMILY CERVIDAE

Rocky Mountain elk (*Cervus elaphus nelsoni*)

Common at higher elevations. They are more reclusive than deer but can be spotted in the ponderosa pine zone and above. In spring and fall, they are often spotted on the former ski resort slopes at Cuchara, near the Dike Trail, and on Boyd Mountain. During summer, small herds of elk head for the meadows above timberline where they can be seen from high trails in the southern Sangres, such as on any slopes near Trinchera Peak. Small herds can also be seen at lower elevations and occasionally cross forest roads in front of cars.

Although usually not aggressive to humans, they are much larger and more unpredictable than deer. Follow these basic rules when you encounter elk. (1) Observe from a distance. (2) Observe from the vehicle if you are in one. On the trail, use binoculars. (3) Cows may attack if you get between them and their offspring. (4) Don't approach bulls at any time. (5) For good observation, stay quiet, speak in whispers, and keep other noises to a minimum.

Rocky Mountain mule deer (*Odocoileus hemionus*)

Mule deer are very common in the region and often seen along roadsides, especially from August through the fall. There may be 700,000 "mulies" statewide. Named for their oversized ears, they have evenly forked antlers and bound away with a stiff-legged gait with their tail held down. They are much less shy than the eastern white-tailed deer, often grazing along the roadsides. Drive carefully at dusk and just after dark; collisions are not uncommon.

Mule deer are 4 to 6 feet long, and a large buck may weight around 400 pounds. The average doe is perhaps half that size. As with white-tailed deer, males grow antlers through the summer and use them to assert dominance and defend their harems during the fall rutting season. Antlers are shed each spring but can only be found shortly afterward. Small rodents gnaw them away as sources of calcium and other nutrients.

Deer are browsers, gaining their nutrients and calories from twigs, leaves, shrubs, and tree branch tips; they may eat crops such as corn and ornamental plants within subdivisions.

White-tailed deer (*Odocoileus virginianus*)

In Colorado, they are mostly reported in or near streamside woodlands of the eastern plains. Bucks are easily distinguished by their single-beamed antlers instead of the even forking of the mule deer. They lift their white tails when running.

Deer and elk disease: Chronic wasting disease (CWD) is a fatal neurological disease that has been found in a few Colorado populations of wild or captive elk, mule deer, and white-tailed deer. As of this writing (2010), there has been no finding that CWD has been transmitted to a human being. But its bovine relative, mad cow disease, can be so transmitted through eating contaminated beef. Also, CWD has been experimentally transmitted to a variety of mammals including primates. For now, the World Health Organization recommends against allowing CWD-contaminated meat to enter the human food system.

CWD is one of the transmissible spongiform encephalopathy (TSE) class of animal diseases. Infection is by abnormal prions, very small protein particles that are smaller than most viruses. Initially, TSE diseases seem not to cause a detectable immune response. With time the infected animal develops mental dementia and loss of weight before inevitable death. The number of possible modes of transmission is unknown, but oral infection by ingesting contaminated materials (e.g., saliva, urine, feces) seems to be one way. The Colorado

Division of Wildlife, in cooperation with federal agencies, is destroying or quarantining all known infected stock and is testing large samples during each harvest season. If you enjoy eating wild game, keep yourself informed of current CWD findings.

PRONGHORN, FAMILY ANTILOCAPRIDAE

Pronghorn antelope (*Antilocapra americana*)

This pretty and engaging animal can be seen bounding or browsing on plains shrubs and on wide open areas of these counties at lower elevations. Pronghorns have a tan coat with white rump patch, are about 4 to 5 feet long, and weigh about 100 pounds. As the fastest American mammals, they can reach 60 mph for short distances.

Pronghorns coexist well with cattle, as they are browsers rather than grazers. Around 50,000 pronghorns are believed to live in Colorado, and in this region they are frequently seen on roads south of La Junta. You are most likely to see one or several around dawn or dusk. They are fun to observe but difficult to approach closely.

BISON, CATTLE, SHEEP, FAMILY BOVIDAE

Bison (*Bison bison*)

Magnificent brown grazing animals, a large bull can weigh as much as 1800 pounds. Native populations are long extirpated from the state. From original remnant populations elsewhere, bison now exist in Colorado occasionally as livestock on ranches. The nearest small herd in a natural setting is at Medano-Zapata Ranch, now owned by the Nature Conservancy. This 100,000-acre tract is just south and west of the entrance to Great Sand Dunes National Park in the San Luis Valley. The herd can be viewed on special tours only (contact The Nature Conservancy; see Resources). If you see bison, never approach them as they are very dangerous.

Bighorn sheep (*Ovis canadensis*) (Fig. 93)

Once widespread throughout the Rockies and somewhat eastward, only scattered populations remain. A population of about 40 animals lives at and above

Fig. 93: Mountain sheep grazing on the north slopes of Mount Trinchera. (Photo by G.N. Keating.)

timberline in the Trinchera Peak area west of Cuchara Pass. Often, they can be seen from the Trinchera approach road/trail that leaves from Blue Lake. Their tracks and droppings are encountered frequently in the Sangres above timberline. Photography yielding good images requires an 800-mm telephoto lens and a tripod. Sheep let you observe but only at a distance.

AMPHIBIANS AND REPTILES

According to the Colorado Herpetological Society, the following species checklist has been documented for both counties. Statewide, on average, about 28% more species are expected but not yet documented for a given county. At this time, Huerfano County is much less known than Las Animas County.

In general, we have too little organized information on local life histories, ecology, and species ranges for this corner of the animal kingdom, as well as for this region. Therefore, I regard this list as provisional for the area covered by this guide. Some specimens are listed even though they are just considered probable.

I am indebted to the Colorado Herpetological Society (http://www.coloherps.org) for providing this checklist and for their work on behalf of knowledge and preservation of this interesting fauna, and to the Colorado Department of Wildlife for the original compilation.

Amphibians: Northern cricket frog (*Acris crepitans*); tiger salamander (*Ambystoma tigrinum*); western toad (*Bufo boreas*); Great Plains toad (*Bufo cognatus*); green toad (*Bufo debilis*); red-spotted toad (*Bufo punctatus*); Woodhouse's toad (*Bufo woodhousii*); Great Plains narrowmouth toad (*Gastrophryne olivacea*); canyon treefrog (*Hyla arenicolor*); western chorus frog (*Pseudacris triseriata*); Plains leopard frog (*Rana blairi*); bullfrog (*Rana catesbeiana*); northern leopard frog (*Rana pipiens*); wood frog (*Rana sylvatica*); Couch's spadefoot (*Scaphiopus couchii*); Plains spadefoot (*Spea bombifrons*); Great Basin spadefoot (*Spea intermontana*); New Mexico spadefoot (*Spea multiplicata*).

Turtles: Spiny softshell (*Apalone spinifera*); snapping turtle (*Chelydra serpentina*); painted turtle (*Chrysemys picta*); yellow mud turtle (*Kinosternon flavescens*); ornate box turtle (*Terrapene ornata*).

Lizards: Triploid Colorado checkered whiptail (*Cnemidophorus neotesselatus*); six-lined racerunner (*Cnemidophorus sexlineatus*); diploid Colorado checkered whiptail (*Cnemidophorus tesselatus*); western whiptail (*Cnemidophorus tigris*); plateau striped whiptail (*Cnemidophorus velox*); collared lizard (*Crotaphytus collaris*); variable skink (*Eumeces gaigeae*); many-lined skink (*Eumeces multivirgatus*); Great Plains skink (*Eumeces obsoletus*); longnose leopard lizard (*Gambelia wislizenii*); lesser earless lizard (*Holbrookia maculata*); Texas horned lizard (*Phrynosoma cornutum*); short horned lizard (*Phrynosoma hernandesi*); round-tail horned lizard (*Phrynosoma modestum*); sagebrush lizard (*Sceloporus graciosus*); desert spiny lizard (*Sceloporus magister*); prairie (or plateau) lizard (*Sceloporus undulatus*); tree lizard (*Urosaurus ornatus*); side-blotched lizard (*Uta stansburiana*).

Snakes: Some people, especially youngsters, are attracted to catching and playing with snakes. Two species listed below (marked with an asterisk [*]), the western or prairie rattlesnake and the massasauga, are classed as "pit vipers" and have poisonous bites. They inject poison through a pair of elongated teeth in their upper jaw. Rattlesnakes are usually not serious problems for people, but pay attention when walking, especially at lower elevations (6000 to 7000 feet). No one should attempt to handle them without training.

Rattlesnakes are named for a rattle borne on the end of the tail that can be used to make a warning sound, although sometimes they strike without warning. The smaller massasauga may reach 26 inches long. Western rattlesnakes may reach 48 inches in our area and have been reportedly found as high as 9500-foot elevation.

Questions arise about aggressiveness and striking distance, which herpetologists say are hard to generalize and depend on circumstances such as the vigor of the snake, whether it is angry or startled, and whether it is bluffing. Normally (whatever that is) strikes are defensive. It especially annoys them to be stepped on or teased. Given a chance, a snake will try to escape. They do not chase people. Rattlesnakes strike from a coiled position, and larger ones can reach a distance of up to two thirds of their body length. Herpetologists have reported that smaller ones can really launch themselves, perhaps up to full length.

What about non-poisonous snake species? Although some are not aggressive when handled, many will attempt to bite. Any kind of snakebite should be considered a septic wound that will become infected unless treated.

Finally, avoid killing snakes. They are important to the ecosystem by keeping down rodent populations and, indirectly, keeping tick-borne and other diseases at low levels.

Glossy snake (*Arizona elegans*); racer (*Coluber constrictor*); western or prairie rattlesnake* (*Crotalus viridis*); ringneck snake (*Diadophis punctatus*); corn snake (*Elaphe guttata*); western hognose snake (*Heterodon nasicus*); night snake (*Hypsiglena torquata*); common kingsnake (*Lampropeltis getula*); milk snake (*Lampropeltis triangulum*); Texas blind snake (*Leptotyphlops dulcis*); smooth green snake (*Liochlorophis vernalis*); coachwhip (*Masticophis flagellum*); striped whipsnake (*Masticophis taeniatus*); northern water snake (*Nerodia sipedon*); gopher snake (*Pituophis catenifer*); long-nosed snake (*Rhinocheilus lecontei*); massasauga* (*Sistrurus catenatus*); ground snake (*Sonora semiannulata*); Plains blackhead snake (*Tantilla nigriceps*); blackneck garter snake (*Thamnophis cyrtopsis*); western terrestrial garter snake (*Thamnophis elegans*); Plains garter snake (*Thamnophis radix*); common garter snake (*Thamnophis sirtalis*); lined snake (*Tropidoclonion lineatum*).

BIRD LIFE

This category presents a coverage problem due to the high mobility of birds. I have taken the conservative approach and listed birds where published range maps include these counties as found in Peterson and Peterson (2001), Sibley (2000), and Tekiela (2001). The family naming sequence is based on Sibley (2000) and the American Ornithologists' Union list (http://www.aou.org). The field guides list bird species that live, breed, or spend a season in the mapped range. As with the other checklists, this list is provisional and will need future revision.

A professionally documented list of birds that spend time in the Spanish Peaks region would be of unknown length, as this kind of detailed study has

Fig. 94 (left): Rufous hummingbird on a fishing rod.

Fig. 95 (below): The accumulation of blue grouse feathers belies the presence of a roost tree. Trail from Wahatoya Canyon toward the Bulls Eye Mine site.

yet to be done. Also, historic ranges do not remain static. The trend toward a milder climate in North America in the past few decades has been shown to affect the breeding and migration ranges of many bird species (Fig. 94).

The other approach, as taken by the Colorado Field Ornithologists group (http://www.coloradocountybirding.com), is to include all birds that have been seen in the region: all confirmed sightings, no matter how infrequent. They maintain regularly updated checklists of birds plus detailed descriptions of good birding areas in both counties. The group lists 278 species observed in Las Animas County and 272 species in Huerfano County. I have included 217 species that are considered residents or regulars in some season (Fig. 95).

LOONS, FAMILY GAVIIDAE

Common loon
Gavia immer, migratory

GREBES, FAMILY PODICIPEDIDAE

Eared grebe
Podiceps nigricollis, summer

Pied-billed grebe
Podilymbus podiceps, all year

Western grebe
Aechmophorus occidentalis, summer

Clark's grebe
Aechmophorus clarkii, summer

PELICANS, FAMILY PELECANIDAE

American white pelican
Pelecanus erythrorhynchos, migratory

CORMORANTS, FAMILY PHALACROCORACIDAE

Double-crested cormorant
Phalacrocorax auritus, migratory

HERONS, FAMILY ARDEIDAE

American bittern
Botaurus lentiginosus, summer

Great blue heron
Ardea herodias, migratory

Snowy egret
Egretta thula, summer

Cattle egret
Bubulcus ibis, summer

Black-crowned night heron
Nycticorax nycticorax, migratory/summer

IBISES, FAMILY THRESKIORNITHIDAE

White-faced ibis
Plegadis chihi, migratory

GEESE, DUCKS, FAMILY ANATIDAE

Canada goose
Branta canadensis, all year

Mallard
Anas platyrhynchos, all year

Gadwall
Anas strepera, all year

Northern pintail
Anas acuta, all year

American wigeon
Anas americana, migratory/winter

Northern shoveler
Anas clypeata, summer

Cinnamon teal
Anas cyanoptera, summer

Blue-winged teal
Anas discors, summer

Green-winged teal
Anas crecca, winter

Canvasback
Aythya valisineria, migratory

Redhead
Aythya americana, summer/all year

Ring-necked duck
Aythya collaris, migratory/winter

Lesser scaup
Aythya affinis, migratory

Common goldeneye
Bucephala clangula, winter

Bufflehead
Bucephala albeola, winter

Common merganser
Mergus merganser, winter

Red-breasted merganser
Mergus serrator, migratory

Ruddy duck
Oxyura jamaicensis, summer

AMERICAN VULTURES, FAMILY CATHARTIDAE

Turkey vulture
Cathartes aura, summer

HARRIERS, FAMILY CIRCINAE

Northern harrier
Circus cyaneus, all year

BIRD HAWKS, FAMILY ACCIPITRIDAE

Sharp-shinned hawk
Accipiter striatus, all year

Cooper's hawk
Accipiter cooperii, all year

Northern goshawk
Accipiter gentilis, winter/all year

BUZZARD HAWKS, FAMILY BUTEONINAE

Swainson's hawk
Buteo swainsoni, summer

Red-tailed hawk
Buteo jamaicensis, all year

Ferruginous hawk
Buteo regalis, all year/winter

Rough-legged hawk
Buteo lagopus, winter

Golden eagle
Aquila chrysaetos, all year

Bald eagle
Haliaeetus leucocephalus, winter

OSPREYS, FAMILY PANDIONIDAE

Osprey
Pandion haliaetus, migratory

FALCONS, FAMILY FALCONIDAE

Merlin
Falco columbarius, winter

American kestrel
Falco sparverius, all year

Prairie falcon
Falco mexicanus, all year

Peregrine falcon
Falco peregrinus, migratory

GROUSE, FAMILY PHASIANIDAE

Ring-necked pheasant
Phasianus colchicus, all year

Blue grouse
Dendragapus obscurus, all year

Wild turkey
Meleagris gallopavo, all year

RAILS, COOTS, FAMILY RALLIDAE

American coot
Fulica americana, summer

Virginia rail
Rallus limicola, summer

Sora
Porzana carolina, summer

CRANES, FAMILY GRUIDAE

Sandhill crane
Grus canadensis, migratory

PLOVERS, FAMILY CHARADRIIDAE

Killdeer
Charadrius vociferus, summer

AVOCETS, FAMILY RECURVIROSTRIDAE

American avocet
Recurvirostra americana, all year

SANDPIPERS, FAMILY SCOLOPACIDAE

Greater yellowlegs
Tringa melanoleuca, migratory

Lesser yellowlegs
Tringa flavipes, migratory

Spotted sandpiper
Actitis macularia, summer

Semipalmated sandpiper
Calidris pusilla, migratory

Least sandpiper
Calidris minutilla, migratory

Long-billed dowitcher
Limnodromus scolopaceus, migratory

Common snipe
Gallinago gallinago, all year

Wilson's phalarope
Phalaropus tricolor, summer

GULLS, TERNS, FAMILY LARIDAE

Bonaparte's gull
Larus philadelphia, migratory

Franklin's gull
Larus pipixcan, migratory

Ring-billed gull
Larus delawarensis, migratory

California gull
Larus californicus, summer

Herring gull
Larus argentatus, migratory

Forster's tern
Sterna forsteri, migratory

Black tern
Chlidonias niger, migratory

PIGEONS, FAMILY COLUMBIDAE

Mourning dove
Zenaida macroura, all year

Rock dove
Columba livia, all year

Band-tailed pigeon
Columba fasciata, summer

CUCKOOS, FAMILY CUCULIDAE

Greater roadrunner
Geococcyx californianus, all year

OWLS, FAMILY STRIGIDAE

Long-eared owl
Asio otus, all year

Short-eared owl
Asio flammeus, winter

Great horned owl
Bubo virginianus, all year

Spotted owl
Strix occidentalis, all year

Northern saw-whet owl
Aegolius acadicus, all year

Burrowing owl
Athene cunicularia, summer

Flammulated owl
Otus flammeolus, summer

Northern pygmy-owl
Glaucidium gnoma, all year

GOATSUCKERS, FAMILY CAPRIMULGIDAE

Common poorwill
Phalaenoptilus nuttallii, summer

Common nighthawk
Chordeiles minor, summer

SWIFTS, FAMILY APODIDAE

White-throated swift
Aeronautes saxatalis, summer

HUMMINGBIRDS, FAMILY TROCHILIDAE

Black-chinned hummingbird
Archilochus alexandri, summer

Broad-tailed hummingbird
Selasphorus platycercus, summer

Rufous hummingbird
Selasphorus rufus, migratory

KINGFISHERS, FAMILY ALCEDINIDAE

Belted kingfisher
Ceryle alcyon, all year

WOODPECKERS, FAMILY PICIDAE

Lewis's woodpecker
Melanerpes lewis, all year

Williamson's sapsucker
Sphyrapicus thyroideus, summer

Red-naped sapsucker
Sphyrapicus nuchalis, summer

Downy woodpecker
Picoides pubescens, all year

Hairy woodpecker
Picoides villosus, all year

Northern flicker
Colaptes auratus, all year

TYRANT FLYCATCHERS, FAMILY TYRANNIDAE

Olive-sided flycatcher
Contopus cooperi, summer/migratory

Western wood-pewee
Contopus sordidulus, summer

Cordilleran flycatcher
Empidonax occidentalis, summer

Willow flycatcher
Empidonax traillii, summer/migratory

Hammond's flycatcher
Empidonax hammondii, summer

Dusky flycatcher
Empidonax oberholseri, summer

Say's phoebe
Sayornis saya, summer

Eastern kingbird
Tyrannus tyrannus, summer

Western kingbird
Tyrannus verticalis, summer

SHRIKES, FAMILY LANIIDAE

Northern shrike
Lanius excubitor, winter

Loggerhead shrike
Lanius ludovicianus, summer

VIREOS, FAMILY VIREONIDAE

Red-eyed vireo
Vireo olivaceus, summer

Warbling vireo
Vireo gilvus, summer

Plumbeous vireo
Vireo plumbeus, summer

Cassin's vireo
Vireo cassinii, migratory

JAYS, FAMILY CORVIDAE

Steller's jay
Cyanocitta stelleri, all year

Western scrub-jay
Aphelocoma californica, all year

Gray jay
Perisoreus canadensis, all year

Pinyon jay
Gymnorhinus cyanocephalus, all year

Clark's nutcracker
Nucifraga columbiana, all year

Black-billed magpie
Pica hudsonia, all year

Common raven
Corvus corax, all year

American crow
Corvus brachyrhynchos, all year

LARKS, FAMILY ALAUDIDAE

Horned lark
Eremophila alpestris, all year

SWALLOWS, FAMILY HIRUNDINIDAE

Purple martin
Progne subis, migratory

Northern rough-winged swallow
Stelgidopteryx serripennis, summer

Bank swallow
Riparia riparia, summer

Violet-green swallow
Tachycineta thalassina, summer

Tree swallow
Tachycineta bicolor, summer

Cliff swallow
Petrochelidon pyrrhonota, summer

Barn swallow
Hirundo rustica, summer

CHICKADEES, TITMICE, FAMILY PARIDAE

Juniper titmouse
Baeolophus ridgwayi, all year

Black-capped chickadee
Poecile atricapilla, all year

Mountain chickadee
Poecile gambeli, all year

BUSHTITS, FAMILY AEGITHALIDAE

Bushtit
Psaltriparus minimus, all year

NUTHATCHES, FAMILY SITTIDAE

Red-breasted nuthatch
Sitta canadensis, all year

White-breasted nuthatch
Sitta carolinensis, all year

Pygmy nuthatch
Sitta pygmaea, all year

CREEPERS, FAMILY CERTHIIDAE

Brown creeper
Certhia americana, all year

WRENS, FAMILY TROGLODYTIDAE

Bewick's wren
Thryomanes bewickii, all year

House wren
Troglodytes aedon, summer

Marsh wren
Cistothorus palustris, all year

Rock wren
Salpinctes obsoletus, summer

Canyon wren
Catherpes mexicanus, all year

DIPPERS, FAMILY CINCLIDAE

American dipper
Cinclus mexicanus, all year

KINGLETS, FAMILY REGULIDAE

Golden-crowned kinglet
Regulus satrapa, all year

Ruby-crowned kinglet
Regulus calendula, summer

GNATCATCHERS, FAMILY SYLVIIDAE

Blue-gray gnatcatcher
Polioptila caerulea, summer

THRUSHES, FAMILY MUSCICAPIDAE

Townsend's solitaire
Myadestes townsendii, all year

Mountain bluebird
Sialia currucoides, all year

Western bluebird
Sialia mexicana, all year

American robin
Turdus migratorius, all year

Veery
Catharus fuscescens, summer

Swainson's thrush
Catharus ustulatus, summer

Hermit thrush
Catharus guttatus, summer

MIMIDS, FAMILY MIMIDAE

Gray catbird
Dumetella carolinensis, summer

Northern mockingbird
Mimus polyglottos, summer/all year

Sage thrasher
Oreoscoptes montanus, summer

STARLINGS, FAMILY STURNIDAE

European starling
Sturnus vulgaris, all year

PIPITS, FAMILY MOTACILLIDAE

American pipit
Anthus rubescens, summer/migratory

WAXWINGS, FAMILY BOMBYCILLIDAE

Cedar waxwing
Bombycilla cedrorum, all year

WOOD WARBLERS, FAMILY PARULIDAE

Orange-crowned warbler
Vermivora celata, migratory/summer

Tennessee warbler
Vermivora peregrina, migratory

Virginia's warbler
Vermivora virginiae, summer

Nashville warbler
Vermivora ruficapilla, migratory

Yellow warbler
Dendroica petechia, summer

Yellow-rumped warbler
Dendroica coronata, summer/migratory

Townsend's warbler
Dendroica townsendii, migratory

Black-and-white warbler
Mniotilta varia, migratory

American redstart
Setophaga ruticilla, migratory

Ovenbird
Seiurus aurocapillus, summer

Northern waterthrush
Seiurus noveboracensis, migratory

MacGillivray's warbler
Oporornis tolmiei, summer

Common yellowthroat
Geothlypis trichas, summer

Wilson's warbler
Wilsonia pusilla, summer/migratory

Yellow-breasted chat
Icteria virens, summer

TANAGERS, FAMILY THRAUPIDAE

Western tanager
Piranga ludoviciana, summer

GROSBEAKS, FAMILY CARDINALIDAE

Black-headed grosbeak
Pheucticus melanocephalus, summer

Blue grosbeak
Guiraca caerulea, summer

Lazuli bunting
Passerina amoena, summer

Indigo bunting
Passerina cyanea, summer

EMBERIZINES, FAMILY EMBERIZIDAE

Spotted towhee
Pipilo maculatus, all year

Green-tailed towhee
Pipilo chlorurus, summer

Canyon towhee
Pipilo fuscus, all year

American tree sparrow
Spizella arborea, winter

Brewer's sparrow
Spizella breweri, summer

Clay-colored sparrow
Spizella pallida, migratory

Chipping sparrow
Spizella passerina, summer

Savannah sparrow
Passerculus sandwichensis, summer

Vesper sparrow
Pooecetes gramineus, summer

Lark bunting
Calamospiza melanocorys, summer

Lark sparrow
Chondestes grammacus, summer

White-throated sparrow
Zonotrichia albicollis, migratory

White-crowned sparrow
Zonotrichia leucophrys, all year

Fox sparrow
Passerella iliaca, summer/migratory

Song sparrow
Melospiza melodia, all year

Lincoln's sparrow
Melospiza lincolnii, summer/migratory

Dark-eyed junco
Junco hyemalis, all year

McCown's longspur
Calcarius mccownii, migratory

Chestnut-collared longspur
Calcarius ornatus, migratory

Lapland longspur
Calcarius lapponicus, winter

ICTERIDS, FAMILY ICTERIDAE

Western meadowlark
Sturnella neglecta, all year

Brown-headed cowbird
Molothrus ater, summer

Yellow-headed blackbird
Xanthocephalus xanthocephalus, summer

Red-winged blackbird
Agelaius phoeniceus, all year

Brewer's blackbird
Euphagus cyanocephalus, all year

Common grackle
Quiscalus quiscula, summer

Great-tailed grackle
Quiscalus mexicanus, summer

Bullock's oriole
Icterus bullockii, summer

FINCHES, FAMILY FRINGILLIDAE

Evening grosbeak
Coccothraustes vespertinus, all year

Pine grosbeak
Pinicola enucleator, all year/winter

Gray-crowned rosy-finch
Leucosticte tephrocotis, winter

Brown-capped rosy-finch
Leucosticte australis, winter/all year

Black rosy-finch
Leucosticte atrata, winter

Cassin's finch
Carpodacus cassinii, all year

House finch
Carpodacus mexicanus, all year

Red crossbill
Loxia curvirostra, all year

Pine siskin
Carduelis pinus, all year

Lesser goldfinch
Carduelis psaltria, summer

American goldfinch
Carduelis tristis, all year

OLD WORLD FINCHES, FAMILY PASSERIDAE

House sparrow
Passer domesticus, all year

RESOURCES

ANNOTATED GUIDE TO THE SPANISH PEAKS LITERATURE

(The wood pulp section. Read so that trees won't have died in vain.)
Both general and technical sources are included.

Alvarez, W. 1997. T. Rex and the Crater of Doom. Princeton University Press, Princeton, NJ. 236 pp. [Sixty-five million years ago, the Mesozoic era and the reign of the giant reptiles came to an abrupt end. This entertaining read explores the current ideas explaining this remarkable event and its aftermath.]

Anderson, D. C. & S. R. Rushforth. 1977. The cryptogam flora of desert soil crusts in southern Utah, U.S.A. Nova Hedwigia 28: 691–729.

Anon. 1999. Pierson Guides Colorado. Recreational Road Atlas. Pierson Graphics, Denver. Spiral bound. [A recreational map guide to roads and features of the state. Deserves to be more widely available. This is one of various types of guides from this publisher.]

Anon. 2008. Reminiscences of Cuchara Valley. Cuchara Hermosa, Cuchara, CO. 222 pp. [A brief history of the Cuchara Valley and of the valley residents, many of whom go back several generations.]

Anon. Colorado Atlas and Gazetteer. DeLorme, Yarmouth, ME. 104 pp. [Published annually. Widely available and accurate. Divides the state into about 100 pages of maps that include topographic data and most obscure roads in all counties.]

Anon. Scenic Highway of Legends. Chambers of Huerfano, La Veta/Cuchara, and Trinidad/Las Animas Counties. 16 pp. [Frequently revised. An illustrated road log for Route 12 between La Veta and Trinidad, giving history and scenery descriptions.]

Arno, S. F. & R. P. Hammerly. 1984. Timberline. Mountain and Arctic Forest Frontiers. The Mountaineers, Seattle. 304 pp. [A lifetime of observations on the ecology and natural history of these environments. Excellent ink drawings by Hammerly. Technical text but mostly accessible to readers with some biological and geographical background. For southern Rockies, see pp. 180–188.]

Ashley, J., S. R. Rushforth & J. R. Johnson. 1985. Soil algae of cryptogamic crusts from the Uintah Basin, Utah, USA. Great Basin Naturalist 45: 432–454. [Technical taxonomic account of soil algae of surface crusts of that region.]

Baldridge, W. S. 2004. Geology of the American Southwest. Cambridge University Press, New York. 280 pp. [A thorough and well-illustrated description of the history and processes through geological time. Previous background would be helpful.]

Barbour, M. G. & W. D. Billings (editors). 1999. North American Terrestrial Vegetation. Cambridge University Press, New York. [Illustrated chapters describe the vegetation and physical geography of ecosystems of the continent.]

Bartlett, R. A. & W. H. Goetzmann. 1982. Exploring the American West, 1803–1879. National Park Service, Department of the Interior, Washington, D.C. 127 pp. [Brief review of all the major American explorations. Just touches this area. Well illustrated.]

Bass, R. 1995. The Lost Grizzlies: A Search for Survivors in the Wilderness of Colorado. Houghton Mifflin, New York. 239 pp. [This well-known outdoor writer joins a group looking for the remnant population of big bears in the San Juan Wilderness. Conclusion: the bears may be there, but solid proof is missing. A good read by an opinionated environmentalist.]

Bates, R. L. & J. A. Jackson (editors). 1984. Dictionary of Geological Terms, Ed. 3. American Geological Institute, Anchor Books, Doubleday, New York. 571 pp. [A standard reference.]

Bell, R. A. & M. R. Sommerfeld. 1987. Algal biomass and primary production within a temperate zone sandstone. Amer. J. Bot. 74: 294–297. [Technical account of rock-dwelling algae.]

Bell, R. A., P. V. Athey & M. R. Sommerfeld. 1988. Distribution of endolithic algae on the Colorado Plateau of Northern Arizona. Southw. Naturalist 33: 315–322. [Technical account of rock-dwelling algae.]

Beshoar, M. 1882. All about Trinidad and Las Animas County, Colorado. Their Histories, Industries, Resources. Times Steam Printing House, Denver. 106+ pp. [A history written by a medical doctor whose practice was in Trinidad.]

Burdett, W. H. 1996. The Roads of Colorado. Shearer Publishing, Fredericksburg, TX. 168 pp. [Very accurate maps that include contour lines.]

Carter, J. L. 1988. Trees and Shrubs of Colorado. Johnson Books, Boulder. 165 pp. [Keys, excellent ink drawings, and descriptions of the woody flora. Brief descriptions of the ecological zones.]

Christofferson, N. C. 2000. Coal Was King. Huerfano County's Mining History. Privately published, La Veta, CO. 114 pp. [Describes the exploitation of the area's coal seams in the early 20th century, as well as the exploitation of the miners.]

Chronic, H. 1980. Roadside Geology of Colorado. Mountain Press, Missoula. 334 pp. [Accessible to beginners, but even more useful if you have some background. Good maps, black-and-white overview photos, and clear road logs. Just touches on this area. Good when driving around Colorado.]

Clyne, R. J. 1999. Coal people. Life in southern Colorado's company towns, 1890–1930. Colorado Hist. 3: 1–121. [Popular article covering area mining history.]

Cronquist, A., A. H. Holmgren, N. H. Holmgren & J. H. Reveal. 1972. Intermountain Flora. Vascular Plants of the Intermountain West, U.S.A. Geological and Botanical History of the Region, Its Plant Geography and a Glossary. The Vascular Cryptogams and the Gymnosperms. New York Botanical Garden (Hafner), Bronx. 270 pp. [Volume 1 of several volumes. It contains much valuable introductory material on geography, ecology, and major plant communities. Not directly germane to this area but good western background.]

Cross, H. E. & J. C. Jochem. 1970. River of Friendship. A Story of Early Cuchara Camps. Nortex Press, Austin. 120 pp. [Stories compiled from longtime residents of the Cuchara camps, dating to 1908. Early settlement and land use history is also detailed.]

Cummings, L. A. (author, compiler, and editor). 1947 (reprinted 1981). History of the Spanish Peaks Ranger District and Surrounding Country. San Isabel National Forest. Region 2. Privately printed typescript. 151 pp. [A valuable historical document, with a fascinating miscellany of reports and summaries written by Cummings, including news story reprints and other reports. Chapters describe forestry, ranching, mining practices, and productivity with some statistics. Also describes game management, trail development, wildlife, and much else.]

Curtis, B. F. 1960. Major geologic features of Colorado. Pp. 1–8 *in* R. J. Weimer & J. D. Haun (editors), Guide to the Geology of Colorado. Geological Association of America. Rocky Mountain Association of Geologists, Denver. [A general account, but the reader should have some technical background.]

Curtis, B. F. (editor). 1975. Cenozoic History of the Southern Rocky Mountains. Mem. Geol. Soc. Amer. 144, Boulder. [Spanish Peaks, p. 81. Technical geological account. The various papers cover the history of all the ranges in this region.]

Donnell, J. R. 1960. Geological Road Logs of Colorado. Rocky Mountain Association of Geologists, Denver. [Written for the public but offers only light coverage of outcrops between Walsenburg and Trinidad along present-day I-25.]

Duffus, R. L. 1930, 1958, 1972. The Santa Fe Trail. University of New Mexico Press, Albuquerque. 283 pp. [Interesting and engagingly written history. The mountain branch of the Santa Fe Trail bordered the eastern margin of the Spanish Peaks.]

Dunmire, W. W. & G. D. Tierney. 1997. Wild Plants and Native Peoples of the Four Corners. Museum of New Mexico Press, Santa Fe. 312 pp. [The plants and how they contribute to the lives of native peoples of the region. A rich source.]

Elmore, F. H. & J. R. Janish. 1976. Shrubs and Trees of the Southwest Uplands. Southwest Parks and Monuments Association, Tucson. 214 pp. [Illustrated guide to the botany and uses of shrubs and trees. Strongest coverage is south and west of here.]

Fassett, J. E. & J. K. Rigby, Jr. 1987. The Cretaceous-Tertiary boundary in the San Juan and Raton Basins, New Mexico and Colorado. Special Paper 209: 1–3; 111–130; 131–150. [Gives description and locality data for the K/T boundary, which was marked by a bolide impact.]

Flora of North America Editorial Committee. 1993. Introduction. *In* Flora of North America North of Mexico, Vol. 1. Oxford University Press, New York. 372 pp. [This volume contains a number of introductory reviews covering geology, past and present climate, physiography, and plant geography, among other useful general topics. More than a dozen subsequent volumes of this continuing series are technical accounts grouped by plant family.]

Flora of North America Editorial Committee. 1993. Flora of North America North of Mexico, Vol. 2. Pteridophytes and Gymnosperms. Oxford University Press, New York. 496 pp. [This series of volumes, a work in progress, provides the most authoritative information for all North American plant families.]

Flores, D. 2001. The Natural West: Environmental History of the Great Plains and Rocky Mountains. University of Oklahoma Press, Norman. 285 pp. [Describes western climate and land and their effect on early people. Prehistoric exploitation of resources was not done sustainably.]

Fralish, J. S. & S. B. Franklin. 2002. Taxonomy and Ecology of Woody Plants in North American Forests. John Wiley, New York. 612 pp. [The only woody plant flora and ecology for the continent. Good combination of ecology and species accounts.]

Garrard, L. H. 1847, 1955. Wah-toy-ah and the Taos Trail. University of Oklahoma Press, Norman. 298 pp. [Written by a young man who traveled this route in its heyday. Garrard had a keen ear for dialect and wrote with great style. It is considered the classic historic text for the region.]

Halfpenny, J. C. 1988. Field Guide to Mammal Tracking in Western North America. Johnson Publishing, Boulder. [Considered the best such guide covering Colorado mammals.]

Hammerson, G. A. 1999. Amphibians and Reptiles in Colorado. University Press of Colorado, Boulder. 484 pp. [Best available field guide covering many aspects of natural history.]

Hart, G. 1977. Free Land for Sale. The Give and Take of Pioneer Prosperity. Leonard Printing, Bartlesville, OK. 99 pp. [Personal reminiscences about homesteading and ranching life in early Huerfano County.]

Herrero, S. 2002. Bear Attacks. Their Causes and Avoidance. Globe Pequot Press, Guilford, CT. 282 pp. [The author is a leading expert on this topic. Here are clearly written case histories and what you need to know about black and grizzly bear behavior. Illustrated.]

Hirleman, N. C. 1981. Historical Map of Las Animas County, Colorado. Self-published, La Veta, CO. [A large folded, hand-drawn map of the county that includes existing features and towns, as well as sites of historical towns and trails. Includes dates and both contemporary and obsolete names. Also includes pen drawings of some historical buildings.]

Hirleman, N. C. 1982. Historical Map of Huerfano County, Colorado. Self-published, La Veta, CO. [See comments under Hirleman (1981).]

Hodge, B. L. 1992. Fort Garland—A Window onto Southwest History. San Luis Valley Historian 24(2): 5–48. [Fort Garland was the U.S. Army presence in the San Luis Valley in the mid- to late-19th century and was a crossroads for troop movements, settlers, and traders. Kit Carson was once post commandant.]

Hopkins, R. L. & L. B. Hopkins. 2000. Hiking Colorado's Geology. The Mountaineers, Seattle. [Good introduction to the nature of the various ranges and peaks, ages and strata. Some stratigraphic diagrams included.]

Howell, J. 2001. Elk reintroduced in Smoky Mountains. Appalachian Trail News Nov/Dec: 16–17. [Although written for an eastern national park, it includes good information on elk behavior and how to observe them.]

Izett, G. A. 1990. The Cretaceous/Tertiary boundary interval, Raton Basin, Colorado and New Mexico and its content of shock-metamorphosed minerals. Evidence relevant to the K/T boundary impact-extinction theory. Special Paper Geol. Soc. Amer. 249. 100 pp. [Technical review of the evidence pointing to the collision of a large asteroid with the earth 65 million years ago.]

Johnson, R. B. 1960. Brief description of the igneous bodies of the Raton Basin region, south central Colorado. In R. J. Weimer & J. D. Haun (editors), Guide to the Geology of Colorado. Geological Society of America, Rocky Mountain Association of Geologists, Colorado Scientific Society, Denver. 310 pp.

Jung, P. P. II, N. W. Bower, M. A. Snyder & D. W. Lehmpuhl. 2001. An application of air sampling canisters: Selective herbivory associated with monoterpene emission from ponderosa pine. Amer. Lab. Feb. 2001: 50–52. [Technical account of methodology for sampling organic gas emissions from trees.]

Kartesz, J. T. & C. A. Meacham. 1999. Synthesis of the North American Flora. Version 1.0, CD ROM. University of North Carolina, Chapel Hill. [This county-level floristic database is useful in developing local flora checklists.]

Keck, F. B. 2001. The JJ Ranch on the Purgatory River. Otero Press, La Junta, CO. 132 pp. [A history of the Jones brothers' ranch in the Purgatory Valley at Nine Mile Bottom between Trinidad and La Junta in the 19th century.]

Kennemer, B. 2001, 2005. Walkin' in a winter wonderland. Signature 2001 Winter Guide. P. 3; Campin' in the cold. Signature 2004–2005 Winter Guide. P. 12. [The author, a longtime area naturalist, writes occasional expert pieces for *The Signature* newspaper and its seasonal visitor guides. In these, he describes essential gear and techniques needed to enjoy the outdoors safely in winter.]

Kessler, R. E. (editor). 1994. Anza's 1779 Comanche Campaign. Diary of Governor Juan Bautista de Anza. San Luis Valley Historian 26; reprinted by R. E. Kessler. [An account of Governor Anza's campaign against the Comanche tribe led by their great war chief Cuerno Verde (Greenhorn). The high peak of the Wet Mountains bears the chief's name. Kessler helped rekindle interest in Governor Anza, who also founded San Francisco. Translated from Spanish by A. B. Thomas.]

Kessler, R. E. 1995. Re-tracing the Old Spanish Trail North Branch. Adobe Village Press, Monte Vista, CO. [Historical account of this San Luis Valley pioneer trail.]

Knox, J. 1983. Mammals of the Northern Great Plains. University of Nebraska Press, Lincoln. 379 pp. [Covers eastern Colorado; minor overlap with this region.]

Knutson, R. M. 1987. Flattened Fauna: A Field Guide to Common Animals of Roads, Streets, and Highways. Ten Speed Press, Chicago. 88 pp. [Good coverage of what is unfortunately the most likely encounter that many people, including biologists, have with many species.]

Lavender, D. 1954. Bent's Fort. Doubleday, Garden City, NY. 450 pp. [A thorough history of this commercial fort near La Junta, CO. It was a major supply and warehousing stop on the Santa Fe Trail for several decades in the mid-19th century.]

Lechleitner, R. R. 1969. Wild Mammals of Colorado: Their Appearance, Habits, Distribution, and Abundance. Pruett, Denver. 254 pp. [Still considered authoritative, although out of print.]

Lindsey, D. A. 1995. Geologic map of the Cuchara Quadrangle, Huerfano County, Colorado. U.S. Geological Survey Misc. Field Studies Map MF-2283. Scale 1:24,000.

Little, E. L. 1971. Atlas of United States Trees, Vol. 1. Conifers and Important Hardwoods. USDA Forest Service Misc. Publ. 1146. U.S. Government Printing Office, Washington, D.C. [An authoritative collection of maps, in six volumes, which shows the ranges of U.S. tree species. Written by America's best-known dendrologist.]

Love, J. D., J. C. Reed & K. L. Pierce. 2003. Creation of the Teton Landscape. A Geological Chronicle of Jackson Hole and the Teton Range, Ed. 2. Grand Teton Natural History Association, Moose, WY. 129 pp. [Although not local, this superbly illustrated and authoritative guide clarifies many processes that pertain to recent Rocky Mountain geological history in general.]

Marino, N. & M. Marino. 1981. Plants of the Alpine Tundra. Rocky Mountain National Historical Association, Estes Park, CO. 65 pp. [Color photos and descriptions of close-ups, habit views, and scenery. Also ink drawings and general natural history of the tundra, somewhat overlapping this area.]

Martinez, W. O. 2001. Anza and Cuerno Verde. Decisive Battle. El Escrito, Pueblo, CO. 114 pp. [A well-researched history of Anza's Cuerno Verde campaign together with a review of

Spanish colonization history in the New World. The focus of new work is a reinterpretation of the final battle site where Cuerno Verde was killed. Martinez makes a persuasive case that Cuerno Verde's death occurred at the St. Charles River, 9 miles north of the present marker west of Colorado City. Genealogists will appreciate publication of the names of soldiers accompanying Anza.]

Matthews, V., K. Kellerlynn & B. Fox (editors). 2003. Messages in Stone. Colorado's Colorful Geology. Colorado Geological Survey, Denver. 157 pp. [Clearly written and very well illustrated. A virtual Geology 101, given that numerous geological phenomena occur within Colorado.]

McGovern, G. S. & L. F. Guttridge. 1972, 1996. The Great Coalfield War. University of Colorado Press, Boulder. [The first and still definitive history of early coal mining in southern Colorado. Presidential candidate George McGovern's doctoral thesis was rendered to good style by his co-author.]

McKinney, D. 2001. Guide to Colorado State Wildlife Areas. Westcliffe Publishers, Boulder. 360 pp. [Lists 240 places with seasons, regulations, and much else, with emphasis on a sportsman's point of view.]

McPhee, J. 1981. Basin and Range. Farrar, Straus and Giroux, New York. 216 pp. [Using field trips and interviews with veteran geologists, this gifted writer interprets the western landscape with great clarity and human interest.]

Murray, J. A. 1985. The Indian Peaks Wilderness Area. A Hiking and Field Guide. Pruett, Boulder. 181 pp. [Guide to a wilderness area just south of Rocky Mountain National Park. Much of the descriptive natural history also pertains to this area.]

Mutel, C. F. & J. C. Emerick. 1984. From Grassland to Glacier. The Natural History of Colorado. Johnson Books, Boulder. 238 pp. [Organized by ecosystem from plains grassland to alpine tundra. Black-and-white photos, ink drawings, checklists of living plants and animals for each ecosystem, and clear descriptions of everything from climatic features to animal habits.]

Nardine, H. 1998. In the Shadows of the Spanish Peaks: A History of Huerfano County, Colorado. Self-published, Walsenburg, CO. 200 pp. [A good review of the settlement and early history.]

Nelson, R. A. 1992. Handbook of Rocky Mountain Plants, Ed. 4. Revised by R. L. Williams. Roberts Rinehart, Niwot, CO. 444 pp. [Technical and illustrated guide to regional plants.]

Noblett, J. B. 1994. A Guide to the Geological History of the Pikes Peak Region. Geology Department, Colorado College, Colorado Springs. 43 pp. [An illustrated and clearly explained introduction to the Rocky Mountains in general as well as the Pikes Peak region.]

O'Hanlon, M. 1999. The Colorado Sangre de Cristo: A Complete Field Guide, Ed. 3. Hungry Gulch Press, Westcliffe, CO. 71 pp. [Trails and basic orientation to the region. Although billed as "the only comprehensive guide," coverage of this neighborhood is light.]

Pearson, M. & J. Fielder. 1994. The Complete Guide to Colorado's Wilderness Areas. Westcliffe Publishers, Boulder. 340 pp. [Accurate and detailed with historical descriptions and fairly good maps. Fielder is a well-known Colorado landscape photographer.]

Peattie, D. C. 1953. A Natural History of Western Trees. Houghton Mifflin, Boston. 751 pp. [Excellent tree lore; too old for authoritative Latin names.]

Peet, R. K. 1978. Latitudinal variation in southern Rocky Mountain forests. J. Biogeogr. 5:

275–289. [An account of life zones and vegetation sampling, including the southern Sangre de Cristos and Spanish Peaks.]

Penn, B. S. & D. A. Lindsey. 1996. Tertiary igneous rocks and Laramide structure and stratigraphy of the Spanish Peaks Region, South Central Colorado. Road log and description from Walsenburg to La Veta (first day) and La Veta to Aguilar (second day). Colorado Geological Survey, Open File Report 96-4: 9. [An informative technical paper written for geologists.]

Peterson, R. T. & V. M. Peterson. 2001. A Field Guide to Western Birds, Ed. 3. Houghton Mifflin, New York. 432 pp. [An updated version of this valuable classic guide.]

Petrides, G. A. & O. Petrides. 2000. Trees of the Rocky Mountains and Intermountain West. Explorer Press, Williamston, MI. 107 pp. [Pocket-sized. Clear descriptions and good nomenclature. Olivia Petrides' drawings are excellent.]

Probst, N. B. 1989. The South Platte Trail. Story of Colorado's Forgotten People. Pruett, Boulder. 244 pp. [Mostly treats areas north of here, but also covers early regional history.]

Roach, G. 1992. Colorado's Fourteeners. From Hikes to Climbs. Fulcrum Publishing, Golden, CO. 249 pp. [This is a thorough guide for those attempting climbs or hiking among Colorado's tallest peaks.]

Robinson, P. 1966. Fossil Mammalia of the Huerfano Formation, Eocene, of Colorado. Peabody Mus. Nat. Hist. Bull. No. 20. 130 pp. [A technical account of mammal fossils as found in various area strata.]

Rockwell, W. 1998 (reprint). The Utes. A Forgotten People. Western Reflections Publishing Co., Ouray, CO. 307 pp. [A readable history of the tribe, covering the Colorado groups and those that occupied this region. Includes a biography of the famous chief Ouray.]

Russell, E. W. B. 1997. People and the Land Through Time: Linking Ecology and History. Yale University Press, New Haven, CT. [Considers the relationship between habitat and the development of human culture.]

Russo, R. 2001. Mountain State Mammals: A Guide to the Mammals of the Rocky Mountain Region. Nature Study Guild, Rochester, NY. 132 pp. [The most recent compilation of mammals of the Rockies.]

Schumacher, B. A. 2004. Paleontological Treasures of Picket Wire Canyonlands, A Glimpse into the Purgatoire River Valley, Comanche Grassland, Southeastern Colorado. Geological Society of America Abstracts and Program, Vol. 36(5): 413. [http://gsa.confex.com/gsa/2004AM/finalprogram/abstract_78805.htm. A brief description of the context of the dinosaur tracks region.]

Scott, G. R. 2001. Historic Trail Map of the Trinidad 1° × 2° Quadrangle, Southern Colorado. U.S. Geological Survey, Washington, D.C. [Detailed, illustrated historical map with lengthy informative accompanying booklet. A better layout would have made the map larger and the pictures smaller.]

Sibley, D. A. 2000. The Sibley Guide to Birds. (National Audubon Society.) Alfred A. Knopf, New York. 545 pp. [The best illustrated guide and best organized. With range maps and much miscellaneous information.]

Sleep, N. H. 2005. Evolution of the continental lithosphere. Annu. Rev. Earth Planet. Sci. 33: 369–393. [Technical paper on the evolution of the earth's outermost 125 miles.]

St. Clair, L. L. 1999. A Color Guidebook to Common Rocky Mountain Lichens. M. L. Bean Life Science Museum, Provo, UT, and San Juan-Rio Grande National Forest, Durango,

CO. 242 pp. [One of the few such regional guides available. Good introduction to lichen study techniques and their importance, with excellent color photographs.]

Snow, N. 2009. Checklist of Vascular Plants of the Southern Rocky Mountain Region. Version 3. 316 pp. [See Colorado Native Plant Society site (http://www.conps.org/plant_lists_keys.html) or (www.botanicus.org/title/b1334416x)].

Snyder, M. A. 1992. Selective herbivory by Abert's squirrels mediated by chemical variability in ponderosa pine. Ecology 73: 1730–1741. [Technical account of the feeding habits of these squirrels and their preference for certain pine chemistry.]

Spackman, S., B. Jennings, J. Coles, C. Dawson, M. Minton, A. Kratz & C. Spurrier. 1997. Colorado Rare Plant Field Guide. Colorado Natural Heritage Program, Fort Collins, CO. [Looseleaf; pages not numbered. A list based on known records of the Colorado flora from herbarium collections.]

Spence, M. L. (editor). 1984. The Expeditions of John Charles Fremont, Vol. 3. Travels from 1848–1854. University of Illinois Press, Urbana and Chicago. [These volumes are a detailed compilation of correspondence and descriptions of the often incomplete journals and maps available for these expeditions.]

Sprugel, D. G. 1989. The relationship of evergreenness, crown architecture, and leaf size. Amer. Naturalist 133: 465–479. [Technical paper comparing relative photosynthetic efficiency of conifers with that of broad-leaved trees.]

Stahle, D. W. 2002. The unsung ancients. Nat. Hist. 2(2): 47–53. [An account of massive and/or very old trees extant on the North American continent.]

Stark, P. 1997. The Old North Trail. Smithsonian July: 54–66. [A re-tracing of a legendary prehistoric trail running from the arctic to southern Colorado.]

Taylor, M. F. 1964. Pioneers of the Picketwire. Trinidad State Junior College, O'Brien Printing and Stationery, Pueblo, CO. 67 pp. [Stories of Nine Mile Bottom and Picketwire Canyon east of Trinidad.]

Taylor, M. F. 1966. Trinidad, Colorado Territory. O'Brien Printing and Stationery, Pueblo, CO. 214 pp. [A general history of the town and environs before statehood.]

Tekiela, S. 2001. Birds of Colorado. Field Guide. Adventure Publishing, Cambridge, MN. 315 pp. [Excellent color photos, range maps for all seasons, life history details, and field notes. Coverage is not comprehensive.]

Thomas, P. 2000. Trees: Their Natural History. Cambridge University Press, New York. 286 pp. [A text that covers all aspects of tree biology and ecology.]

Vigil, K. 2001. Historic battleground. Puebloan pinpoints 1779 site of Cuerno Verde-Anza battle. The Pueblo Chieftain, July 16, 2001. [Reviews the work of Wilfred Martinez: Anza and Cuerno Verde: Decisive Battle. Based on Anza's diary, Martinez believes the battlefield is a gully of the St. Charles River near the junction of Water Barrel and Burnt Mill roads, north of the state-recognized site.]

Voynick, S. M. 1994. Colorado Rockhounding. Mountain Press, Missoula. 371 pp. [No mention of Huerfano County. Otherwise a thorough account of minerals and their characteristics, as well as where to find them.]

Watts, M. T. & T. Watts. 1974. Desert Tree Finder. Nature Study Guild, Berkeley. 61 pp. [A pocket guide with good maps and drawings and not overly technical keys.]

Weber, W. A. 1972. Rocky Mountain Flora. Colorado Associated University Press, Boulder. 438 pp. [An authoritative technical field manual of the flora.]

Weimer, R. J. & J. D. Haun (editors). 1960. Guide to the Geology of Colorado. Geological Society of America, Rocky Mountain Association of Geologists, Colorado Scientific

Society, Denver. 310 pp. [A collection of well-illustrated technical papers covering most eras and periods of the state's geological history. Includes field trip logs and extensive bibliography.]

Willard, B. E. & C. O. Harris. 1976. Alpine Wildflowers of the Rocky Mountains, Ed. revised. Rocky Mountain Nature Association, Estes Park, CO. 24 pp. [Despite their identical titles, this and the next listing have largely different illustrations and text.]

Willard, B. E. & M. T. Smithson. (n.d.). Alpine Wildflowers of the Rocky Mountains. Rocky Mountain Nature Association, Estes Park, CO. 39 pp. [Short, well-illustrated pocket guide.]

Williams, R. L. & R. A. Nelson. 1992. Handbook of Rocky Mountain Plants, Ed. 4. Roberts Rinehart Publishers, Niwot, CO. 444 pp. [A thorough and easy-to-use guide to the flora. Authorities for Latin names are lacking.]

Wilson, D. R. 1990. Colorado. Historical Tour Guide. Crossroads Communications, Carpentersville, IL. 440 pp. [Well-illustrated guide with much miscellaneous information. Fairly comprehensive listings of numerous historical spots in the state. Lists several stops in Walsenburg, Ludlow, and Trinidad.]

Zwinger, A. H. & B. E. Willard. 1972. Land Above the Trees. Harper & Row, New York. 489 pp. [Excellent exposition about geology, geomorphology, and biology of the tundra.]

WEB SITES

The rate at which web sites come and go prevents up-to-the-minute accuracy. According to http://www.archive.org, the average life of an extant or unchanged web page is 100 days. Therefore, this list cannot pretend to be comprehensive. Use of search engines will produce much additional information. For factual information, I favor academic, scientific, and agency web sites as being more reliable than private ones where you cannot evaluate the source.

Animals: Amphibians and reptiles: http://www.coloherps.org [Colorado Herpetological Society maintains checklists and other useful information.]

Animals: Mammals: http://www.wildlife.state.co.us/WildlifeSpecies/Profiles [Much basic information, and the only checklist for Colorado.]

Animals: Mammals: http://www.mnh.si.edu/mna [This site encourages authoritative searches on North American mammals, by species or by state.]

Birds: http://www.birds.cornell.edu/AllAboutBirds/BirdGuide [A searchable database with bird photos, classification, natural history, and miscellany.]

Colorado Natural Resources Department: http://www.dnr.state.co.us [News items, articles on wildlife, online license sales, and much more.]

Colorado State University Cooperative Extension: http://www.ext.colostate.edu [Miscellaneous natural resources data, information on pest outbreaks; links to other natural resource–based agencies.]

County information: http://www.huerfano.us and http://www.huerfanocountychamberofcom merce.com [Basic demographic data. Huerfano County Homepage (as of summer 2010).]

Forest pest outbreaks: Colorado State Cooperative Extension Service: http://www.ext.colo state.edu/menu_insect.html [Good for information on forest pest outbreaks and much else regarding Colorado natural resources. Also has links to other natural resource agencies.]

Geology: http://www.emporia.edu/earthsci/aberjame.htm [J. S. Aber and his students have studied the Spanish Peaks region for years. Good illustrations including aerial photography; authoritative text on glaciation and bedrock geology.]

National Forests: http://www.fs.fed.us/r2/psicc [Much information on the history and features of western national forests.]

Natural History: http://cumuseum.colorado.edu/ [Checklists and other information based on the collections at the University of Colorado.]

Photography: http://www.photo.net [This is a huge discussion group and general presentation site for all things photographic. Many other photographic topical sites exist.]

Plants: http://www.unco.edu/nhs/biology/environment/herbarium [University of Northern Colorado herbarium. Interactive keys to southern Colorado flora can be downloaded.]

Plants: http://www.plants.usda.gov [Good profiles of most species of local plants with descriptions, maps, and color photos.]

Soil organisms: http://www.soilcrust.org [Has a technical bibliography covering cryptobiotic crusts.]

Soil organisms: Belknap, J. 2001. Cryptobiotic soils: Holding the place in place. Available online at http://www.geochange.er.usgs.gov/sw/impacts/biology/crypto/ [A web site explaining the nature of cryptobiotic soil crusts and their value in holding soil water and nutrients.]

Wildlife: http://www.ndis.nrel.colostate.edu/wildlife.asp [Much wildlife information for the state.]

AGENCIES AND ORGANIZATIONS

AGENCIES

Bureau of Land Management
Colorado State Office
2850 Youngfield Street
Lakewood, CO 80215
Tel: 303-239-3600
http://www.co.blm.gov

La Jara Field Office (BLM, USFS)
[Lower San Luis Valley]
15571 County Road T5
La Jara, CO 81140
Tel: 719-274-8971; Fax: 719-274-6301

Royal Gorge Field Office (BLM, USFS)
[Spanish Peaks Region]
3028 East Main Street
Cañon City, CO 81212
Tel: 719-269-8500; Fax: 719-269-8599

Colorado State University (CSU) Cooperative Extension Service
For Huerfano County:
928 Russell Avenue
Walsenburg, CO 81089
Tel: 719-738-2170

For Las Animas County:
2200 North Linden Avenue
Trinidad, CO 81082
Tel: 719-846-6881

Colorado State Parks
Southeast Region Office
4255 Sinton Road
Colorado Springs, CO 80907
Tel: 719-227-5250; Fax: 719-227-5264
http://www.parks.state.co.us

Lathrop State Park
70 County Road 502
Walsenburg, CO 81089
Tel: 719-738-2376

Trinidad State Park
32610 Highway 12
Trinidad, CO 81082
Tel: 719-846-6951

Colorado Department of Natural Resources
Colorado Division of Wildlife (CDOW)
State Wildlife Areas
1313 Sherman Street, Suite 719
Denver, CO 80203
Tel: 800-244-5613
Shop@DOW [for guide to state wildlife areas]
http://www.wildlife.state.co.us

Regional Service Center
4255 Sinton Road
Colorado Springs, CO 80907
Tel: 719-227-5200

CDOW Area Office
600 Reservoir Road
Pueblo, CO 81005
Tel: 719-561-5300

When planning a site visit to an SWA:
Huerfano County Sheriff's Department
Tel: 719-738-1600
Las Animas County Sheriff's Department
Tel: 719-846-2211

Comanche National Grasslands, USDA Forest Service
[Administrator of Picketwire Canyonlands; Dinosaur Track Site]
Pike and San Isabel National Forest, Timpas Unit
1420 East 3rd Street
La Junta, CO 81050
Tel: 719-384-2181
www.fs.fed.us/r2/psicc
[Contact them for tours and access to the site south of La Junta.]

Pike and San Isabel National Forest
USDA, Forest Service
1920 Valley Drive
Pueblo, CO 81008
Tel: 719-545-8737

U.S. Geological Survey Information Services
Book and Map Sales
Box 25286
Denver Federal Center
Denver, CO 80225
Tel: 303-202-4700; 800-435-7627

ORGANIZATIONS

The Nature Conservancy of Colorado
2424 Spruce Street
Boulder, CO 80302
Tel: 303-444-2950
http://www.nature.org/colorado

Medano-Zapata Ranch
The Nature Conservancy
San Luis Valley Program
5303 Highway 150
Mosca, CO 81146
Tel: 719-378-2356 (tour information)

CHAMBERS OF COMMERCE

Huerfano County Chamber of Commerce
P.O. Box 493
Walsenburg, CO 81089
Tel: 719-738-1065

La Veta-Cuchara Chamber of Commerce
P.O. Box 32
La Veta, CO 81055
Tel: 719-742-3676

Trinidad/Las Animas County Chamber of Commerce
309 Nevada Street
Trinidad, CO 81082
Tel: 719-846-9285

LIBRARIES

La Veta Public Library
310 S. Main Street
P.O. Box 28
La Veta, CO 81055
Tel: 719-742-3572

Spanish Peaks Library District
415 Walsen Avenue
Walsenburg, CO 81089
Tel: 719-738-2774

Trinidad: Carnegie Public Library
202 N. Animas Street
Trinidad, CO 81082
Tel: 719-846-6841

MUSEUMS

Francisco Fort Museum
308 South Main Street
La Veta, CO 81055
Tel: 719-742-5501
http://www.spanishpeakscountry.com/FortFrancisco.aspx

Walsenburg Mining Museum
101 East 5th Street
P.O. Box 134
Walsenburg, CO 81089
Tel: 719-738-1992
http://www.spanishpeakscountry.com/WalsenburgMiningMuseum.aspx

APPENDIX 1

PREPARING FOR OUTINGS

AT THE END of a beautiful day you've seen happy hikers (maybe you!) coming off the trail with no equipment, dressed lightly and wearing flimsy sandals. So why should one worry about special preparation or equipment?

Consider this scenario, a true story from the summer of 2002. On a late July day, the morning began warm and clear. A young woman, her two young daughters, and their pet dog had set out to walk the Baker Creek Trail. Two hours later, a former wilderness guide and his friend also took that trail. As the day wore on, the sky darkened and a gentle rain quickly became cold and wind driven, changing to hail and then sleet. The temperature dropped 30 degrees.

The two men happened upon the mother and her daughters, who had stopped and were shivering severely. The daughters were becoming hypothermic, and their mother was nearly hysterical. They had missed a turn and were lost. Previously they had fed the dog the extra food and had no insulation or rain gear. The young men lent them extra jackets, gave them food, and accompanied them back to their car, often carrying the daughters. By grace and luck, a happy ending.

This cautionary tale is not meant to frighten anyone. Many of us are glad to drive across the continent to experience these gorgeous mountains in any mood. Sensible preparation makes it possible. Personally, I enjoy the atmosphere of "inclement" weather. The issue is that weather conditions can change quickly in mountain environments, and your ability to adapt will not just save the day but allow you to enjoy the experience in comfort. Only you can build up experience, so this section is dedicated to helping you prepare and have the right supplies at hand.

The following comments and checklists are intended as a primer for those planning half-day or all-day excursions on trails, or into the wilderness at higher elevations. You don't need to spend a lot of money all at once. As an aside, though, in most western states, hikers and bird watchers are now bringing in more tourist dollars than hunters and fishermen.

The "in-tents" experience of overnight backpacking and winter hiking is beyond the scope of this guide, but the literature list includes some useful sources (see Resources). Your first such trips should be in the company of experienced backcountry campers.

CONDITIONING

Unquestionably, being in good shape contributes directly to your enjoyment of nature. No matter what your general physical condition, it can be improved by a regular workout routine. It need not cause stress or pain; a good conditioning program should be planned to yield gradual improvement over a period of months and years.

If you arrive here from low elevations, breathing may be a little labored for a few days. At 9000 feet, available oxygen is about 70% of that at sea level. For most people in good health, it will take a few days for your blood's oxygen-carrying capacity to adapt. Arriving here in good general condition is very helpful.

Trainers agree that discouragement with exercise is usually due to pushing too hard in the beginning. Begin your routine with modest goals. Reduce weights or pace if you experience strain or pain. Be patient and accept gradual increases in fitness. Those with physical problems should get the advice of their physician or a physical therapist as to the best and safest pathway to conditioning. Older people, especially, should do gentle stretching before weight training.

Ideally, you should begin your program 6 to 12 months before you arrive. To prepare for hiking or walking at higher elevations, treadmills, step machines, and exercise bikes are all useful and can be used in any weather. When exercising outdoors, walking, jogging, running, or cycling produces excellent results.

If you arrive here from lower elevations, allow several days or longer before undertaking a strenuous walk, especially if you experience unpleasant symptoms of altitude adjustment. Some individuals' physiology resists adapting quickly to high elevation. They may feel sick or headachy or lose appetite as their body struggles for oxygen. Listen to your body. Don't attempt activities that seem foolishly strenuous. Begin with shorter walks before attempting longer trails.

WALKING

Entire books have been written on this seemingly obvious topic. Posture, gait, and foot anatomy vary widely and can be evaluated by a physical therapist when required. Pain in the arch or various foot bones can often be fixed with a well-designed orthotic device prescribed by a therapist or podiatrist.

GUIDE PACE

A common occurrence among novice hikers is running out of breath or experiencing pain in leg muscles 5 or 10 minutes into a trail climb. For years, some mountain guides have recommended the "guide pace." This is a walking strategy originally developed for getting out-of-shape clients up mountains such as Grand Teton as they carry a heavy pack.

Basically, when carrying a pack, beginning hikers are advised to place one foot just in front of the other and match this pace to the rhythm of their breathing. This mincing pace works well. Lengthen your stride as you get your second wind. Keep the rhythm of your pace and breathing synchronized. When you find this groove, you will be pleased with your progress as you put miles behind you.

FOOTWEAR

Good footwear is the first consideration when walking on trails or cross country. Good quality running or cross-training shoes are adequate for shorter walks on well-maintained trails. While some experienced hikers prefer such lightweight footwear, or even sandals with heavy soles, the need for ankle protection dictates 7-inch-high hiking boots when traveling on rougher or longer trails.

Boots come in a myriad of contemporary styles, but basically there are three weights. Lightweight boots with lots of exposed fabric are fine for day hiking. Medium-weight boots have sturdier counter and ankle support and are recommended when carrying a pack of 20% to 30% of your body weight. The heaviest expedition boots provide the necessary support when carrying a large pack, especially for off-trail or wilderness settings. Wearing expedition boots for day trips will cost you more energy and tire you sooner.

You should buy hiking boots in an outfitting store where a knowledgeable salesperson can fit you competently. Describe your experience to the salesperson and what you intend to be doing, and bring heavy socks to use when testing for fit. Poorly fitting cheap boots bought at a discount store will yield big savings in price but have the potential of ruining your outdoor experience. In this category, price is a good guide to satisfaction. Finally, do not select a brand on a friend's recommendation alone. Try several brands in your price range, as proper fit is a personal thing and brands differ in subtle but important ways.

THE 10 ESSENTIALS

The following list is recommended by many sources for those planning all-day hikes. It should be especially heeded by those traveling into unfamiliar or wilderness landscapes. They are as follows:

1. Map and compass, and know how to use them.
2. Rain protection and extra clothing to handle potentially adverse conditions.
3. Extra food, more than you think you'll need.
4. Water, more than you think you'll need.
5. Flashlight with good batteries.
6. Fire starters, matches in a waterproof case. Fire encouragement, such as heat tabs, candle stubs, and chunks of paraffin-impregnated sawdust, is especially useful in wet weather. Note that under drought conditions, fires may be banned. Heed local authorities' directives.
7. Sun protection for head and eyes, and sunscreen. The importance of these increases with elevation.
8. First aid kit. Review its contents before your hike, and know how to use them.
9. Knife. A pocket knife with a couple of tools is preferable to a sheath knife.
10. Bivy bag, an emergency sleeping bag. It can be a space blanket, which is a thin, Mylar plastic sheet with a reflective surface. Even a pair of large garbage bags can be fashioned into a body heat–preserving cover.

ANNOTATED CHECKLIST

These suggestions pertain to plans for day hikes, and only summer-season gear is outlined here. Winter outings and camping needs are outside the scope of this guide.

CLOTHING

Remember that in alpine areas one must be prepared for more intense sun exposure than at sea level. Also, even in summer, a variety of weather conditions can occur even in a single day.

General Tips: Garments should be chosen to allow a complete range of movement as is typical of today's looser-fitting styles. Also note that cotton is a poor choice if there is any chance of getting wet from rain or sweat. Wet cotton becomes clammy, heavy, and uncomfortable and may cause chills if the temperature drops.

Underwear: Whatever is usual may work. However, there are well-designed undergarments for both men and women, made of synthetics (nylon, polypropylene, or polyester) designed to wick moisture away from the skin. They are quite comfortable. Only outdoor stores carry this type of gear, and it is worth the extra money.

Hiking Pants: Cotton jeans are a poor choice. They are hot on warm days, cold on cold days, and heavy and awful when wet. Good synthetics or cotton/synthetic blends are available that are lightweight and wick moisture away from you. Convertible nylon or nylon-blend pants with zip-off legs work very well. They are light, stand away from the skin, and are fast drying when one encounters a passing rain shower. Shorts, although popular, provide no defense against the sun, a scratchy brush, or a change to cold weather.

Shirts: There are lots of choices here. The best are synthetic or synthetic-blend styles with two breast pockets for carrying glasses, notes, and so on. Modern hiking shirts have nice features and feel comfortable over a range of weather conditions. Some now come with sunblock protection built into the fabric, which is worth the price if you use them often.

Weather Protection/Insulation: Rain protection comes in many forms, ranging from expensive Gore-Tex–type laminates to inexpensive nylon or plastic ponchos. For insulation, polar fleece vests, pullovers, or jackets are excellent. As with wool, they remain warm when wet, but they are lighter than wool. Layering is the concept. It is better to carry two lighter items than one heavy one so as to adjust to your comfort level.

Hat: A mandatory piece of gear as part of your sun-protection scheme. A hat with an all-around brim is best. Cotton or straw works well in warm weather. A lightweight acrylic or fleece watch cap gives good protection against high cold winds when above timberline.

Bandana: A good all-purpose tool: emergency hat, sling, towel, bandage, etc. Carry more than one.

CARRYING YOUR GEAR

Size your day pack generously, larger than just roomy enough to carry lunch. There should be space for rain gear and insulation items, as well as any cameras or special interest items you carry. Well-designed hip belts and sternum straps are very effective in distributing weight for maximum comfort. Lightweight waterproof pack covers are a worthy accessory for rain protection, especially if you need to protect a camera. For short trips, hip or fanny packs that fasten around the waist are good for carrying water and lunch.

WHAT TO CARRY

"Weight is the enemy" is a backpacker's maxim. Even for day hikes, carry only what is essential for your safety and enjoyment of nature. In the end, only experience will inform you.

Water: In arid areas, 4 quarts per person per day. If your route includes strenuous climbing, avoid carbonated soft drinks that day and the previous day. For a day or more, extra carbon dioxide reduces your blood's oxygen-carrying capacity.

Lunch: Ample supply, twice as much as you normally eat on a sedentary day. Walking is labor, and labor runs on food. Fatigue that you suspect comes from being out of shape may just be a need for more fuel.

Snacks: A bag of gorp or trail mix, a mixture of dried fruit pieces, nuts, and small candies. This is good emergency food. Some people like snacking; others never do.

Sun Protection: Use sunblock cream of at least SPF 30, or preferably SPF 45, and lip balm of similar ratings. Select a product that has UVA as well as UVB protection. Ultraviolet rays in the "A" spectrum penetrate the skin deeply and cause damage that can result later in skin cancer. Your hat and clothing are also significant parts of your sunscreen. Some find that long trousers and long sleeves are more comfortable than skin slathered with lots of cream.

Toilet Paper: When used in nature, get away from the trail and stream. Cover waste and paper well with dirt, rocks, or downed wood.

First Aid Kit: A personal kit should include Band-Aids, butterfly adhesive strips, moleskin, adhesive tape, small scissors, several sizes of gauze bandages, 3-inch-wide ace bandage, burn cream, antiseptic cream, small tweezers for splinter removal, pain pills, and topical pain killer.

In larger parties, one person would be well advised to carry a larger kit that includes splints and more gear. In order to avoid flailing around and wasting time in the field, review basic first aid in advance.

Fire Starter: For an emergency warming fire, use matches or a small butane lighter. For cooking or atmosphere in the wilderness, avoid lighting wood-fueled fires on the ground. This antique concept has caused too many wildfires in recent years. Also, the search for wood in some popular areas has produced a scoured and worn landscape. If you must have a ground fire, be fanatical in extinguishing it before leaving. If you wish to cook, use one of the well-designed backpacking stoves that are available, but use these with caution too. Unless you are backpacking, carrying ready-to-eat food is fine.

Map, Compass, and GPS: Despite the fact that you studied the route in advance, unexpected landmarks can cause confusion in the field. Understand how to read your map and compass. The latest GPS are a pleasure to use, but they have a learning curve. They can trace your movement and list your map coordinates and elevation. They do so by triangulating signals received from an array of navigation satellites. They are no substitute for maps unless you have loaded the appropriate map into its memory. I emphasize: be observant. Cultivate your "map sense," a feeling for where you are based on paying attention to your surroundings as you travel. Such a mind-set serves you well.

Sit-upon: A small piece of closed foam padding is useful when resting on damp ground.

Notebook and Pencil: All sorts of unexpected or interesting observations cry out to be recorded.

Cord: A small hank of parachute shroud or nylon cord can come in handy.

Natural History Interests: Field guides to plants or animals, a hand lens on a neck cord, a camera, a lightweight tripod—anything you are willing to carry.

Walking Stick or Matched Pair: Some hikers never carry these. A stick can be useful support for distributing weight and effort, as when crossing a stream or traveling cross country. Some hikers consider pairs of high-tech adjustable metal walking sticks to be essential. But there have been criticisms in the trail community; the sharp pointed sticks are said to tear up trails unnecessarily. Adjustable metal walking sticks or monopods are effective but might act as lightning rods in high country. A wooden or bamboo stick with a rubber tip is best.

APPENDIX 2

NOTES ON COLORADO TRESPASS LAW

NOTE: The following summary is not a legal opinion. In case of question or challenge, ask your lawyer, do your own research, or seek advice from the appropriate agency. For a useful discussion, see http://www.co.blm.gov/lsra/huntfaqs.htm. Long-time residents of Colorado tell of wandering all over private lands in their youth and being perfectly welcome to do so. Now, however, we live in an era of gated communities and many other signs of increased tension between visitors and landowners.

1. Generally: Observe all signage and postings, do not litter, and leave gates as you found them. Show respect and be a good neighbor.

2. Private property: Landowner permission is required to enter private lands. Light to severe penalties can be assessed by courts. *Land does not have to be posted* in order for trespass law to be enforced.

3. Public roads under state or local jurisdictions are open to public travel. They cross private land by legal easement, but the adjacent private land cannot be entered without permission. Local public roads can be closed by an agency because of public health, safety, or wildlife management issues.

4. Identifying public roads can be a problem. Public roads are those that appear on state maps and those with numbers on atlas pages. U.S. Geological Survey topographic quadrangles show all visible roads regardless of ownership. Sometimes signage is missing. If in doubt, ask at police, state, or federal offices for clarification.

5. Public lands are open to public access unless specifically closed for safety or wildlife management purposes. Grazing lessees or permittees on public land have no authority to restrict or control public access. Note, however, that livestock are private property and they must not be harassed or their movements interfered with in any way.

6. Trust lands, identified in places by roadside signs, were created when Colorado became a state. Often, they are not open to the public but are managed to produce income for local schools and public institutions.

7. Bureau of Land Management (BLM) lands are not common in this area. Often occurring in small parcels surrounded by private property, they are accessible only if a public road easement exists, or by permission of the adjacent landowner.

8. If access problems arise, back off from confrontation. Report the problem to the agency with jurisdiction, such as local law enforcement, the Colorado Division of Wildlife, a BLM field office, or a state land office.

APPENDIX 3

PHOTOGRAPHY IN THE MOUNTAINS

THIS SECTION contains tips for people who use 35-mm or other film cameras with speed and aperture adjustments, and perhaps a choice of lenses. Consider the suggestions as starting points. Those who understand their systems often produce stunning results when they "break the rules." If you use digital point-and-shoot cameras, your results will most likely be satisfactory anyway or can be redone on the spot. Indulge the rest of use who still love film and enjoy fiddling with equipment.

METERING

The quality of light at high elevations is different from what you encounter on the coasts or in the Midwest. With clear skies at, say, above 9000 feet, there appears to be about one more zone, or a one-stop wider dynamic range. Some photographers suggest closing down one more stop than the meter indicates, or using the "sunny 22 rule"; that is, *f*/22 at the shutter speed closest to the film's exposure index. In a hazier atmosphere, you can estimate using the usual "sunny 16 rule." Another way is to take a reading on a pocket-sized 18% gray card, available at photo stores. As the pros know, the only way to know for sure is to test your equipment and favorite film under the conditions you favor. This is impossible, of course, for the visitor with limited time.

As camera shutters and meters age, they may drift away from original speeds. Rather than getting expensive service, simply adjust your film speed ratings. If you have little experience in mountain photography and want to be assured of getting good results, bracket your exposures (expose as indicated, plus one stop overexposed and one stop underexposed), and keep notes.

If you are photographing an especially bright scene with somewhat harsh lighting, your meter might give erratic instructions as you compose the image.

The following tip can help: Light in a bright blue sky, metered at its darkest point at right angles to the sun, should equal the value of an 18% gray card. Also called *Zone V*, it is the average exposure for the scene.

LENSES

Wide Angle: Many mountain photographers find they get much use from short focal length prime lenses; for example, 28 or 35 mm. The juxtaposition of a flowered, detailed foreground and peaks in the background makes an interesting composition.

Zoom: When I carry only one camera and lens, my choice is a 28- to 85-mm zoom. Not all zoom lenses are as good as prime lenses, but I have found one that is sharp and reliable and provides good contrast. Zooming also allows you to frame a scene without having to walk around a lot to find the best composition. This is especially useful if you are traveling with a group.

Telephoto: These lenses are not especially helpful if your goal is to document mountain scenery but are essential when photographing wildlife. Wildlife photographers recommend an 800-mm lens for this purpose, and certainly no shorter than 500 mm. A tripod is mandatory. A rule of thumb for sharp, handheld exposures is that the shutter speed should be at least as fast as the focal length of the lens (e.g., a 500-mm lens would require a shutter speed of at least 1/500th of a second).

CLOSE-UP WORK

Photographing wildflowers is easier on slightly overcast days, which give a more natural, even lighting. In my opinion, modern color films are at their best in less bright, evenly lit subjects. On very bright days, I sometimes carry a small white umbrella, obtainable at camera shops, which kills strong shadows that otherwise might ruin a nice close-up. White and yellow flowers, especially, tend to have a bleached-out look when photographed in full sun.

FILTERS

Remember that filter effects vary with lens, film type, and processing. You have to experiment. Cheap ones can degrade image quality. Expensive ones are multicoated and yield sharper images.

UV or Skylight Filter: A good idea, not just to protect your lens but also to minimize overexposure. Skylight filter 1A will slightly warm the tone of the scene, and 1B will further warm the scene.

Warming Filter: This type, called 81A, 81B, or 81C in order of increasing effect, will remove the bluish cast found in many well-lit scenes. Many outdoor photographers routinely use these instead of the 1A and 1B types. Some use the 85 series filters to provide the look of early morning or late-afternoon light.

Polaroid Filter: These are often considered quite useful for sharpening colors, especially in flatly lit scenes. I no longer use one at high elevation. In the mountains, Polaroids produce a fake, gaudy look and dark, cobalt-blue skies. They are fine for the occasional dramatic shot. Color films already exaggerate colors, which is why I prefer to capture wildflowers on light overcast days.

Filters for Black-and-White Work:

• Yellow darkens the sky (absorbs blue) and provides an easier-to-print negative of a scene.

• Light green affects the sky as does a yellow filter and also lightens foliage.

• Orange provides an even darker sky.

• Red creates a dramatic scene with black sky, reminiscent, for example, of Ansel Adams' Mount McKinley photos.

INDEX, GENERAL TOPICS

Many geologic terms are found in the geologic glossary on page 64. Most names of organisms are indexed separately.

Y

INDEX OF PLANT AND ANIMAL NAMES

A

C

E

F

T

X

Y

Z